Thermodynamics of Magnetizing Materials and Superconductors

Thermodynamics of Magnetizing Materials and Superconductors

Vladimir Kozhevnikov

CRC Press
Taylor & Francis Group
Boca Raton London New York

CRC Press is an imprint of the
Taylor & Francis Group, an **informa** business

CRC Press
Taylor & Francis Group
6000 Broken Sound Parkway NW, Suite 300
Boca Raton, FL 33487-2742

First issued in paperback 2020

ISBN-13: 978-1-138-49993-5 (hbk)
ISBN-13: 978-0-367-78801-8 (pbk)

Library of Congress Cataloging-in-Publication Data

Names: Kozhevnikov, Vladimir (Retired physics professor), author.
Title: Thermodynamics of magnetizing materials and superconductors / Vladimir Kozhevnikov.
Description: Boca Raton, FL : CRC Press, Taylor & Francis Group, [2019] | Includes bibliographical references and index.
Identifiers: LCCN 2019007144| ISBN 9781138499935 (hbk ; alk. paper) | ISBN 1138499935 (hbk ; alk. paper) | ISBN 9780429266478 (ebk) | ISBN 0429266472 (ebk)
Subjects: LCSH: Magnetic fields. | Magnetic materials. | Superconductors. | Thermodynamics.
Classification: LCC QC754.2.M3 K69 2019 | DDC 538/.3--dc23
LC record available at https://lccn.loc.gov/2019007144

Visit the Taylor & Francis Web site at
http://www.taylorandfrancis.com

and the CRC Press Web site at
http://www.crcpress.com

To the memory of my parents,
Klara Yakovlevna Grossman
and
Fedor Mikhailovich Kozhevnikov

Contents

Preface

Historically, thermodynamics was formed as the science of heat and temperature and the associated work and energy. However, due to the electronic structure of matter and the ubiquitous presence of a magnetic field, magnetic properties of material media are on an equal footing with their heat counterparts. Nevertheless, thermodynamic properties caused by the magnetic field are only scarcely addressed in textbooks, as a short section or an appendix, which cannot be sufficient for in-depth understanding of a far-from-trivial subject of the magnetic properties. Moreover, the magnetic properties considered in textbooks on thermodynamics, condensed matter physics and superconductivity are practically always related (however, this is not always specified) to the simplest geometry of a long cylindrical body in parallel field; whereas in the majority of contemporary research and applications, a specimen/field configuration is close to what is called transverse geometry: infinite plate in a perpendicular field. The magnetic properties of bodies in a crucial manner depend on the geometry of the body/field configuration, which makes correct formulae for one geometry wrong for the other. Unfortunately, this circumstance, indicated in the classical treatises by Maxwell, is practically ignored in the majority of modern textbooks. The suggested monograph aims to reclaim this situation.

Each chapter of the book is accompanied by problems, so it can be used as a supplemental textbook for courses on thermodynamics, electrodynamics, physics of condensed matter and superconductivity. Respectively, the book is intended for students and researchers specializing in the indicated areas of physics, as well as for anyone else interested in physics of magnetism and superconductivity. In the latter case, it is assumed that the reader is well familiar with the standard course of general physics.

I am grateful to Andreas Suter, Vladimir Kresin, Mikhail Raikh, Orest Symko and Dietrich Belitz for reading and commenting on the manuscript. Special thanks to Chris Van Haesendonck without whom this book would never have been written. Last but not least, I am indebted to my wife Yelena Spivak for infinite patience and support at all stages of this project.

Vladimir Kozhevnikov
May 2019

List of Symbols

A   vector potential

B   magnetic induction

$\mathbf{B}_0$ induction of the applied magnetic field

B_c induction of the thermodynamic critical field (used in SI unit system)

B_I induction of the applied field at the transition from the Meissner state to the intermediate state in type-I superconductors (used in SI unit system)

B_n, B_t normal and tangential components of the magnetic induction, respectively

$b_{\parallel}$, $b_{\perp}$ reduced parallel and perpendicular components of the magnetic induction, respectively

C Curie constant

C_H, C_B heat capacities of a specimen at constant H and B, respectively

C_s, C_n heat capacities of a specimen in superconducting and normal state, respectively

c speed of light

D period of a laminar flux structure

D_n, D_s width of the normal and superconducting laminae, respectively

E   electric field

e   microscopic electric field

e electron charge

E_c' condensation energy of a specimen

e_c condensation energy density

e_{kin} kinetic energy of electrons per unit volume

E_m magnetic energy of a specimen

$F(T, V, \mathbf{B})$, $\widehat{F}(T, V, \mathbf{H})$, $\widetilde{F}(T, V, \mathbf{H}_0)$ different forms of a specimen free energy; $\widetilde{F}(T, V, \mathbf{H}_0)$ is a total free energy.

$f(T, V, \mathbf{B})$, $\widehat{f}(T, V, \mathbf{H})$, $\widetilde{f}(T, V, \mathbf{H}_0)$ different forms of free energy density

f   force acting on the unit area of an interface between two media in the magnetic field

$G(T, P)$ Gibbs thermodynamic potential

$g_t(T, P)$ density of the Gibbs thermodynamic potential

g   linear density of a surface current

$\mathbf{g}_c$ linear density of a conduction (transport) surface current

H   magnetic field intensity

$\mathbf{H}_0$ intensity of the applied magnetic field

$h_{0\|}$, $h_{0\perp}$ parallel and perpendicular components of the applied magnetic field intensity in reduced units

H_c thermodynamic critical magnetic field

H_{c1} lower critical magnetic field

H_{c2} upper critical magnetic field

H_{c3} critical magnetic field of nucleation of superconductivity

H_{cr} critical magnetic field of transition from the superconducting to the normal state

H_d demagnetizing magnetic field

H_{ext} external magnetic field near the specimen

H_I magnetic field of the transition from the Meissner state to the intermediate state

H_n, H_t normal and tangential components of the magnetic field intensity, respectively

h   microscopic magnetic field

I   magnetization

j   mean bulk density of a current

$\mathbf{j}_c, \mathbf{j}_b, \mathbf{j}_p$ mean bulk density of a conduction (transport), bound and persistent currents, respectively.

k_B Boltzmann constant

L_h healing length

$\mathbf{L}_m$ angular momentum of electrons in a single molecule

M   magnetic moment of a specimen

$\mathbf{M}_m$ magnetic moment of a single molecule

m electron mass

$\mathbf{n}$ unit vector normal to the specimen surface

n number density

n_s number density of superconducting electrons

n_v vortex number density

$\mathbf{o}$ Larmor angular velocity

P pressure

Q latent heat

S entropy

S_s and S_n entropy of a specimen in superconducting and normal state, respectively

S_m, S_{lat} magnetic and lattice parts of entropy, respectively

s entropy per unit volume of a specimen

T temperature

T_D Debye temperature

T_c critical temperature of the superconducting transition

T_m kinetic energy of electrons in one molecule

t time

U internal energy of a specimen

$u(s, \mathbf{B})$, $\widehat{u}(s, \mathbf{H})$, $\widetilde{u}(s, \mathbf{H}_0)$ different forms of internal energy density

V volume

$\mathbf{v}_{mic}$ velocity of microscopic motion of electrons

W work

γ gyromagnetic ratio

δ domain-wall energy parameter

η demagnetizing factor

κ Ginzburg-Landau parameter

λ magnetic penetration depth

μ_0 permeability of free space (constant of SI unit system)

μ, μ_r magnetic permeability and relative magnetic permeability in *cgs* and SI unit systems, respectively

μ_{ch} chemical potential
ξ coherence length
ρ_n, ρ_s volume fractions of the normal and superconducting components, respectively
σ surface density of electric charges
Φ magnetic flux
Φ_0 superconducting magnetic flux quantum
χ, χ^* magnetic susceptibility in *cgs* and SI unit systems,
respectively
ψ Ginzburg-Landau order parameter
Ψ magnetic scalar potential
Ω number of accessible states

Abbreviations

AMMS   averaged model of the mixed state
e.m.f.   electromotive force
GL   Ginzburg-Landau (theory, parameter, etc.)
FC   field cooled (specimen)
IS   intermediate state
LMTF   laminar model of the intermediate state in a
tilted field
MS   mixed state
N   normal state and phase
PL   Peierls-London model
RRR   residual resistivity ratio
S   superconducting state and phase
ZFC   zero-field cooled (specimen)

Introduction

Thermodynamics is a branch of physical science studying macroscopic properties of systems of a large number of identical constituents, such as molecules, in an equilibrium state also referred to as a state of thermodynamic equilibrium. This is the state uniquely determined by a set of parameters called parameters of state or thermodynamic variables. Therefore, properties of the equilibrium state, referred to as thermodynamic properties, do not depend on the system history, i.e., on the path by which the system was brought at its present state. Examples of parameters of state include temperature, volume, electric potential, and many others. Parameters of state can be vector quantities. In such case, thermodynamic properties can depend on the system (we will also use the terms "body" and "specimen") shape and on its orientation with respect to these parameters. Examples of the vectorial thermodynamic variables are static electric and magnetic fields. In this book we will be concerned with the properties caused by the latter.

Thermodynamics is based on the most solidly established laws of nature: the law of energy conservation and the law of maximal entropy of a macroscopic system in the equilibrium state. For that reason, compliance with the laws of thermodynamics is a necessary condition of validity of any physical result whether it is theoretical or experimental. One of the advantages of thermodynamics is that it allows one to predict and/or explain macroscopic properties of a system even if microscopic processes underlying these properties are not well understood.

Owing to the electronic structure of atoms, a specimen of any material in the magnetic field possesses a magnetic moment $\mathbf{M}$ (this and other quantities mentioned in this section are defined later in the text). For a vast majority of materials $\mathbf{M}$ is an induced property, implying that it vanishes in absence of an applied static or dc (direct current) magnetic field $\mathbf{H}_0$. The latter is the field set by the magnet and hence controlled in the experiment; this is the field away from the magnetized specimen or the field which would be in a space occupied by the specimen if it is absent. Thermodynamic properties of materials with an induced magnetic moment, referred to as *magnetizing* and *magnetizable* materials, constitute a subject of this book. Thus, we will first discuss properties of para- and diamagnetics (Ch. 2), whereas properties of *magnetic* materials, i.e., materials possessing spontaneous magnetization, such as ferromagnetics, will be rarely mentioned[1].

We remind the reader that paramagnetic materials, or simply paramagnetics, are materials whose induced magnetic moment $\mathbf{M}$ is positive with respect to the vector $\mathbf{H}_0$, i.e., $\mathbf{M} \cdot \mathbf{H}_0 > 0$. The reverse is always true for diamagnetics. Normally, magnetism in magnetizing materials is fairly weak. In particular, magnetic permeability μ only slightly differs from unity: typically, it is $1 - O(-5)$ in diamagnetics and $1 + O(-4)$ in paramagnetics [1].

A hallmark of superconductivity phenomenon is a magnetic moment induced by an applied magnetic dc field. Often superconductors are referred to as perfect diamagnetics. This is indeed true for specimens in the Meissner state, defined as a state with a magnetic induction $\mathbf{B} = 0$ over the whole specimen volume. Therefore, it can be viewed as a state with "perfect" diamagnetic permeability $\mu = 0$. However, it is not quite true for superconductors in other equilibrium

---

[1] There are excellent accounts of properties of the magnetic materials, e.g., "Ferromagnetism" by R. M. Bozorth, "The theory of Magnetism" by D. C. Mattis, "Micromagnetics" by F. W. Brown, Jr., "Introduction to Magnetism and Magnetic Materials" by D. Jiles, and other.

superconducting states, e.g., in the mixed state (MS) of type-II superconductors or in the intermediate state (IS) of type-I superconductors. The induction B in such states is inhomogeneous, and at constant temperature the specimens can be considered as diamagnetics with a field-dependent average or effective permeability $0 < \bar{\mu} < 1$. Correspondingly, at a fixed field $\bar{\mu}$ in these states depends on temperature.

The magnetic moment induced in a pure, i.e., pinning free, specimen of any (type-I or type-II) superconductor in any superconducting state at given temperature and pressure is uniquely determined by the applied field. Therefore, $\mathbf{M}$ in a superconductor, like in the "normal" magnetizing materials, is a thermodynamic property. One of the quantitative differences between conventional diamagnetics (such as bismuth, one of the materials with which Faraday discovered the diamagnetism phenomenon [2]) and superconductors is that a spatial scale of the induced circulating currents in the latter is much larger than that in the former. Correspondingly, the induced magnetic moment in superconductors is much greater (up to 5 orders of magnitude!) than that in normal diamagnetic specimens of the same volume and shape. Pure superconductors meet the definition of the magnetizing materials and therefore a large portion of this book (Chapter 3) is devoted to thermodynamic properties of superconductors.

Apart from clarification of the physical meaning of some notions and related mathematical formulae, the main motivation in writing this book is the fact that magnetic properties of matter considered in most textbooks are related to properties of specimens of cylindrical geometry (long cylinders in parallel field), whereas in the majority of contemporary experiments and calculations the specimen/field configuration is close to so-called transverse geometry, i.e., infinite plates or slabs in a perpendicular field. Properties of specimens of this geometry significantly (even drastically in case of superconductors) differ from those of the cylindrical geometry. Correspondingly, magnetic properties of specimens of non-cylindrical geometries are the main focus of this book.

In Chapter 1 we consider elements of magnetostatics of the magnetizing materials and superconductors, necessary for consequent discussion of their thermodynamic properties. The content of this chapter does not imply substitution of textbooks on electrodynamics and superconductivity, so many formulas are presented without derivations but with emphasis on their physical meaning instead.

As a rule (unfortunately, it is not always possible) we use capital characters for quantities related to an entire specimen and lowercase symbols for the quantities related to its unit volume. By default, formulas are given in cgs units; all (but most obvious) formulas in the text are duplicated in SI units labeled (SI). Developed by Gauss with contributions of Maxwell and Thomson (Lord Kelvin) the cgs unit system has well-known advantages for electrodynamics stemming from the same dimension of the electric and magnetic fields in this system. Also, it is important that cgs units are used in all classical treatises on electrodynamics and superconductivity. Therefore we believe that the use of cgs units will help the reader to read original monographs and papers by great physicists, which is always an extremely useful and enjoyable experience[2].

---

[2] An extended account of different aspects of systems of units is available in "Bible of Units" by N. Carron, arXiv:1506.01951 (2015).

CHAPTER 1

Elements of Magnetostatics in Magnetizing Media

The magnetic moment and other magnetic properties of materials are caused by microscopic electric currents circulating inside them (recall that intrinsic magnetic moment of electron associated with its spin angular momentum is also attributed to circulating currents [3, 4]).

The magnetic moment (or dipole magnetic moment) of a body is defined as [5, 6]

$$\mathbf{M} = \frac{1}{2c} \int_V (\mathbf{r} \times \mathbf{j}) dV \qquad (1.1)$$

$$\mathbf{M} = \frac{1}{2} \int_V (\mathbf{r} \times \mathbf{j}) dV, \quad \text{(SI)}$$

where $\mathbf{r}$ is the radius-vector from an origin somewhere inside the body to a physically infinitesimal element[1] of its volume

---

[1]Following Lorentz, physically infinitesimal elements of volumes, surfaces and lines are called such elements which are much smaller than macroscopic inhomogeneities, and at the same time much larger than microscopic inhomogeneities of the medium and the field (in a condensed matter those are of the order of molecular size).

dV in which the mean density of microscopic currents is **j** and the integral is taken over the whole volume of the body V; and c is the electrodynamic constant of the *cgs* unit system equal to speed of light. The mean current density $\mathbf{j} = e\overline{n}\mathbf{v}_{mic}$, where e, n and $\mathbf{v}_{mic}$ are charge, number density and velocity of microscopic motion of electrons, respectively.

1.1 DIELECTRICS

In insulators or dielectrics, electrons are bound to molecules. Accordingly, microscopic currents are referred to as bound currents. Denoting the mean density of these currents $\mathbf{j}_b$, Eq. (1.1) is rewritten as

$$\mathbf{M} = \frac{1}{2c} \int_V (\mathbf{r} \times \mathbf{j}_b) dV = \int_V \mathbf{I} dV, \qquad (1.2)$$

where we introduced *magnetization* **I**, which is magnetic moment density or the moment per unit volume.

As seen from (1.2), **I** is determined by the bound current $\mathbf{j}_b$ and therefore it should be possible to express $\mathbf{j}_b$ in terms of **I**. Indeed, as one can show (see, e.g., [3, 7, 8]), the relationship between $\mathbf{j}_b$ an **I** is

$$\mathbf{j}_b = c\nabla \times \mathbf{I} \qquad (1.3)$$

$$\mathbf{j}_b = \nabla \times \mathbf{I}. \quad \text{(SI)}$$

On first glance, Eq. (1.3) may look rather strange because it yields zero bound current in uniformly magnetized bodies. But if it is so, then, according to (1.2), the magnetic moment of such bodies and hence magnetization are zero as well! Oddly enough, $\mathbf{j}_b$ in these bodies is indeed zero but **M** and **I** are not. Let us see how it happens.

The bound current in Eq. (1.3) is the average of microscopic currents in the volume element dV of the specimen *bulk*, which in case of uniform magnetization *must* be zero due to symmetry. However, any body possesses a surface near which magnetization abruptly (over a length of the order of a radius of circulating currents) *changes* either down to zero, if the body is in free space (vacuum), or to another value, if the body is submerged in some medium. This near-surface change of magnetization results in appearance of a *surface current*. Note the peculiar type of this current: it is formed by electrons bound to molecules and therefore there is no real charges traveling along the surface. However, it is easy to show (see, e.g., [3]) that the magnetic moment produced by this current is exactly the same as that caused by a continuous current running along the surface.

The surface current is characterized by its linear density **g**, defined as a current per unit length of the body boundary perpendicular to the current. The vector **g** (see, e.g., [3, 8] for derivation) is

$$\mathbf{g} = c(\mathbf{I}_1 - \mathbf{I}_2) \times \mathbf{n} = \frac{c}{4\pi}(\mathbf{B}_1 - \mathbf{B}_2) \times \mathbf{n} \qquad (1.4)$$

$$\mathbf{g} = (\mathbf{I}_1 - \mathbf{I}_2) \times \mathbf{n} = \frac{1}{\mu_0}(\mathbf{B}_1 - \mathbf{B}_2) \times \mathbf{n}, \quad \text{(SI)},$$

where $\mathbf{I}_1$ and $\mathbf{B}_1$ are magnetization and induction (definition of **B** is given below) at inner side of the body boundary, respectively; $\mathbf{I}_2$ and $\mathbf{B}_2$ are the same quantities at the outer side (in surrounding medium); **n** is a unit vector perpendicular to the surface and directed outward and μ_0 is a constant of the SI unit system equal $4\pi \cdot 10^{-7} N/A^2$ and referred to as permeability of free space. The last expression in (1.4) follows from the fact that **g** is parallel to the surface since the bound current can never run through (perpendicular to) it.

The first part of Eq. (1.2) accounts only for the bulk bound current $\mathbf{j}_b$. Therefore, it is incomplete and should be supplemented by a term reflecting the near-boundary change of magnetization. On the other hand, the second part of Eq. (1.2) is completely correct because the total magnetic moment of a body can always be presented as a sum of magnetic moments of its volume elements $d\mathbf{M} = \mathbf{I}dV$, which in magnetizing materials are always parallel to each other.

Following Faraday, the magnetic field inside matter is called *magnetic induction* or magnetic flux density and is denoted by $\mathbf{B}$. This is an average value of the microscopic magnetic fields $\mathbf{h}$ obeying the Maxwell equations for free space [9]. Like magnetization, at steady-state conditions the induction is also an equilibrium property of magnetizing materials. As it was pointed out already by Faraday (see [10]), $\mathbf{B}$ is the field available for measurements. At the time of Faraday and Maxwell, one could only speculate about measuring the field inside materials. Nowadays it can be indeed measured, e.g., by means of muon spin rotation (μSR) spectroscopy.

The Maxwell equations for $\mathbf{h}$ are

$$\nabla \cdot \mathbf{h} = 0 \qquad (1.5)$$

and

$$\nabla \times \mathbf{h} = \frac{1}{c}\frac{\partial \mathbf{e}}{\partial t} + \frac{4\pi}{c}en\mathbf{v}_{mic} \qquad (1.6)$$

$$\nabla \times \mathbf{h} = \frac{1}{c^2}\frac{\partial \mathbf{e}}{\partial t} + \mu_0 en\mathbf{v}_{mic} \quad (\text{SI}),$$

where $\mathbf{e}$ is a microscopic electric field (not to be confused with the electron charge e).

Upon averaging, the term with the time derivative in (1.6) vanishes because the mean electric field $\overline{\mathbf{e}}$ is assumed absent. On the other hand, the average microscopic current density in non-conducting media is $\mathbf{j}_b \equiv e\overline{n\mathbf{v}}_{mic} = c\nabla \times \mathbf{I}$ (see Eq. (1.3)). Therefore, Eqs. (1.5) and (1.6), after averaging, become

$$\nabla \cdot \mathbf{B} = 0 \qquad (1.7)$$

and

$$\nabla \times \mathbf{B} = 4\pi\nabla \times \mathbf{I} \qquad (1.8)$$

$$\nabla \times \mathbf{B} = \mu_0 \nabla \times \mathbf{I}. \quad (\text{SI})$$

Introducing

$$\mathbf{H} \equiv \mathbf{B} - 4\pi\mathbf{I}, \qquad (1.9)$$

$$\mathbf{H} \equiv \frac{\mathbf{B}}{\mu_0} - \mathbf{I}, \quad (\text{SI})$$

we rewrite Eq. (1.8) as

$$\nabla \times \mathbf{H} = 0. \qquad (1.10)$$

$\mathbf{H}$ is another equilibrium characteristic of the magnetic field in a matter referred to as magnetic and magnetizing force [2, 10], magnetic field intensity [3], magnetic field [7], auxiliary field [8], thermodynamic field [11], Maxwell field [12], H-field [13], and others.

As seen from (1.10), in non-conducting media (or, in general, in absence of a transport current) curl $\mathbf{H}$ is zero due to definition of $\mathbf{H}$ (1.9). Therefore, in such media, $\mathbf{H}$ represents a potential (vortex-free) field [3] which can be written as

$$\mathbf{H} = -\nabla\Psi, \qquad (1.11)$$

where Ψ is the magnetic scalar potential obeying the Laplace equation

$$\nabla^2\Psi = 0.$$

This kind of description of the magnetic field is used in the Poisson theory of magnetism based on Coulomb law for so-called magnetic charges or monopoles [14, 10]. Eq. (1.10) justifies the validity of the formulae obtained from the notion of the scalar potential in this (physically inconsistent) theory. It should be stressed that curl of the genuine magnetic field (Eq.(1.6)) is *not zero*, meaning that $\mathbf{h}$ and hence $\mathbf{B}$ is a solenoidal (vortex) field, consistently describing by a vector potential $\mathbf{A}$ defined as

$$\nabla \times \mathbf{A} = \mathbf{B}. \qquad (1.12)$$

In the Gaussian or *cgs* unit system H has the same dimension as B ($[Length]^{-1/2} \times [Mass]^{1/2} \times [Time]^{-1} = cm^{-1/2}g^{1/2}s^{-1}$). However the units of B and H carry different names: gauss (G) and oersted (Oe), respectively. In the SI system, B and H have both different dimensions ($[Mass] \times [Time]^{-2} \times [Current] = kg \cdot s^{-2}A$ and $[Current] \times [Length]^{-1} = A/m$, respectively) and names (tesla (T) for B and no name for H).

These and other issues related to unit system and terminology can lead to the impression that B and H are different magnetic fields simultaneously existing inside materials. Actually, however, these are different characteristics of a single concept of the magnetic field used for description of the magnetism phenomenon in classical electrodynamics. It is well seen, for instance, from the fact, that in order to calculate the work δW needed for an infinitesimal change of the induction $\delta \mathbf{B}$ in a volume V_F occupied by the field, one needs to know the field intensity $\mathbf{H}$ in this volume [15]:

$$\delta W = \frac{1}{4\pi} \int_{V_F} \mathbf{H} \, \delta \mathbf{B} \, dV \qquad (1.13)$$

$$\delta W = \int_{V_F} \mathbf{H} \, \delta \mathbf{B} \, dV \qquad \text{(SI)}.$$

Therefore the knowledge of both the $\mathbf{B}$ and $\mathbf{H}$ fields is evenly important for describing magnetic properties of materials. In free space outside of any specimen (where $\mathbf{I} = 0$, but the field can be different from $\mathbf{H}_0$ due to magnetic moment of the specimen) $\mathbf{B}$ and $\mathbf{H}$ are identical (or differ by a constant factor μ_0 in SI units). Correspondingly, away from the specimen, where the field due to the magnetized specimen vanishes, $\mathbf{B} = \mathbf{H} = \mathbf{H}_0$. Inside the specimen $\mathbf{B}$ and/or $\mathbf{H}$ differ from $\mathbf{H}_0$ and must be *carefully distinguished* from each other [10]. It is important that, apart from dependence on the specimen material, the differences between $\mathbf{B}$, $\mathbf{H}$ and $\mathbf{H}_0$ depend on the specimen geometry and on the orientation of the applied field. For instance, if the specimen is an infinite slab in parallel field $\mathbf{H}_0$, then everywhere outside this specimen $\mathbf{B} = \mathbf{H} = \mathbf{H}_0$; inside the slab $\mathbf{H}$ remains the same as $\mathbf{H}_0$, but $\mathbf{B}$ is different: in paramagnetics $B > H_0$ and in diamagnetics $B < H_0$. However, if $\mathbf{H}_0$ is perpendicular to the same slab, then $\mathbf{B} = \mathbf{H}_0$ and $H < H_0$ in paramagnetics and $H > H_0$ in diamagnetics. Below we will consider this issue in more details.

To complete the system of equations (1.7) and (1.10), these equations must be supplemented by a relationship between and by boundary conditions for the B and H fields.

In many magnetizing materials (see, e.g., [16] for details) B and H are linearly related, i.e.,

$$\mathbf{B} = \mu\mathbf{H}, \tag{1.14}$$

where μ is already mentioned *magnetic permeability* defined by this formula; μ is intrinsic property of the magnetizing materials (excluding superconductors). In *cgs* system μ is dimensionless quantity; in SI system Eq. (1.14) is rewritten as

$$\mathbf{B} = \mu_r\mu_0\mathbf{H}, \qquad \text{(SI)}$$

where μ_r is the same as μ in *cgs* system and is referred to as relative magnetic permeability. Materials with the field independent μ are called linear magnetizing materials.

Another intrinsic property, referred to as (bulk) *magnetic susceptibility* or coefficient of induced magnetization (per unit volume) $\chi = (\mu - 1)/4\pi$, is defined by

$$\mathbf{I} = \chi\mathbf{H}. \tag{1.15}$$

In SI units the magnetic susceptibility is $\chi* \equiv \mathbf{I}/\mathbf{H} = \mu_r - 1$.

Hence, in paramagnetics χ is positive and it is negative in diamagnetics.

Apart from that, in paramagnetics χ depends on temperature. In many paramagnetics χ is inversely proportional to

T (Curie Law) and therefore the magnetic part of the specimen entropy S_m (entropy caused by disorder in orientation of molecular magnetic moments) decreases with decreasing temperature (the lower temperature, the less disorder in the ordering of molecular moments in the magnetic field). This leads to so-called magnetocaloric effect, i.e., to heat release at magnetization and heat absorption at demagnetization of paramagnetics. The magnetocaloric effect is used to achieve very low temperatures and in other applications [17, 18]. This effect is considered in problems 2.8-2.10 after Chapter 2.

Contrarily, in diamagnetics χ is temperature independent[2]. This means that the induced molecular magnetic moments in diamagnetics are completely ordered and therefore $S_m = 0$.

Calculation of χ is one of the main tasks of condensed matter and molecular physics [21, 20]. Note, that although classical Langevin theory and quantum mechanics yield similar formulas for χ in dia- and paramagnetics, rigorous application of the classical theory yields $\chi = 0$ [19], indicating that magnetization is a quantum phenomenon [20, 21, 22].

It is worth noting that according to (1.15) I is proportional to H, but not to the average microscopic field B. Historically, definition (1.15) appeared in the Poisson theory and it remains in force in the classical Faraday-Maxwell theory [10, 7]. That the field in (1.15) may not be B can be seen from the fact that magnetization is caused by interaction of molecular currents with an average magnetic field acting on each molecule from its neighbors. The latter apparently does not include the field from the molecule itself, whereas the induction B does include it.

There are different opinions about what field should be used in (1.15) (see, e.g., [3, 13, 20]). The problem is that H is an unmeasurable quantity because any probe (including muons in μSR) measures average field in its vicinity (provided that distortion of the field by the probe is negligible),

---

[2]it is not always so in superconductors.

which is B by definition. And also, in normal materials μ differs very little from unity. Therefore, difference between B and H in these materials is mostly of academic interest. However, in superconductors (a) in the Meissner state (where $B = 0$, but $H \neq 0$) the concept of the magnetic permeability makes sense only if $\mathbf{I}$ is proportional to $\mathbf{H}$, and (b) in the intermediate and mixed superconducting states, where the specimens are split for domains of the normal and supercon- ducting phases, H throughout a specimen equals to B (or B/μ_0 in SI units) in the normal domains and the latter can be and was measured. Results of these measurements support the classical definition of χ given by (1.15). We will briefly consider these results later.

Boundary condition for the normal component of the in- duction B_n follows from always valid Eq. (1.7) stemming from the absence of magnetic monopoles. It yields (see, e.g., [7, 6] for derivation)

$$B_{n1} = B_{n2}, \tag{1.16}$$

where indexes 1 and 2 refer to the induction on the inner and outer sides of the body boundary, respectively.

The condition (1.16) implies that lines of the induction are endless (no beginning, no end) continuous lines. Most often the induction lines form closed loops about currents, which include the lines coming from and leaving to infinity. Note that, at definite combinations of the currents, the B- lines may be infinite but located in a confined space, e.g., they can make an endless toroidal helix (see [3] for details)[3].

Another important content of Eq. (1.16) is that it reflexes the law of the magnetic flux conservation, stating that the magnetic flux (flux of the induction $\mathbf{B}$) passing through any closed surface is always zero. In other words, the magnetic

---

[3]Such a current combination is used in a tokamak, which is the first and, as of today, the most perspective project of a fusion reactor em- ployed, in particular, in the International Thermonuclear Experimen- tal Reactor (ITER). The tokomak was conceptually developed by Igor Tamm and his former PhD student Andrei Sakharov.

flux entering a closed volume (for example, a specimen) is equal to the flux exiting from it.

In its turn, Eq. (1.10) yields the boundary condition for the tangential (parallel to the surface) component H_t, which is (see, e.g., [3, 6] for derivation)

$$H_{t1} = H_{t2}, \qquad (1.17)$$

where meaning of indexes 1 and 2 is the same as that in (1.16).

It is important to underscore that condition (1.17) follows from Eq. (1.10) and therefore it is valid in the absence of the transport or conduction current. The surface current made by circulating microscopic currents does not alter the condition (1.17) since these currents are always compensated. Moreover, even in the presence of the transport current Eq. (1.17) holds provided that the bulk density of this current $\mathbf{j}_c$ (see Eq. (1.34) below) is finite. However, if the transport current is concentrated in a very thin layer at the boundary (for example, the current in a solenoid made of a thin wire wound on a cylindrical specimen, see problems 1.4 and 1.5), then the condition (1.17) takes the form [3, 6]

$$(\mathbf{H}_1 - \mathbf{H}_2) \times \mathbf{n} = \frac{4\pi}{c}\mathbf{g}_c \qquad (1.17a)$$

$$(\mathbf{H}_1 - \mathbf{H}_2) \times \mathbf{n} = \mathbf{g}_c, \qquad \text{(SI)}$$

or in the scalar form

$$H_{t1} - H_{t2} = \frac{4\pi}{c}g_c$$

$$H_{t1} - H_{t2} = g_c, \qquad \text{(SI)}$$

where $\mathbf{g}_c$ is the linear density of the *surface transport current*[4] running in direction perpendicular to H_t.

---

[4]In a multiple connected superconducting specimen g_c can be a persistent current encircling an opening with a trapped magnetic flux.

Contrarily to H_t, the tangential component of induction B_t at the boundary undergoes a discontinuity regardless of presence or absence of the transport current. Condition for B_t for insulators was already given in (1.4). It can be rewritten as

$$B_{t1} - B_{t2} = \frac{4\pi}{c}g \qquad (1.18)$$

$$B_{t1} - B_{t2} = \mu_0 g, \qquad \text{(SI)}$$

In conductors with a transport surface current, g is the linear density of a total surface current. In the absence of the transport current, g is composed of the microscopic currents circulating near the surface.

The physical reason why in the absence of the surface transport current $\Delta B_t \equiv B_{t1} - B_{t2} \neq 0$ while $H_{t1} = H_{t2}$ is well seen from Eq. (1.4): ΔB_t is due to difference in magnetizations of the contacting media. Recall that magnetization is the sum of microscopic magnetic momenta which in magnetizing materials are parallel and therefore can not compensate each other.

The normal component of H at the boundary is also discontinuous. Assuming that the specimen is in free space ($I_2 = 0$ or $\mu_2 = 1$), one finds

$$H_{n2} - H_{n1} = B_{n2} - (B_{n1} - 4\pi I_{n1}) = 4\pi I_{n1} \qquad (1.19)$$

$$H_{n2} - H_{n1} = \frac{B_{n2}}{\mu_0} - (\frac{B_{n1}}{\mu_0} - I_{n1}) = I_{n1}. \qquad \text{(SI)}$$

This condition is similar to that for the electrostatic field $\mathbf{E}$ in metals and for the normal component of $\mathbf{E}$ in polarized dielectrics $E_{n1} - E_{n2} = 4\pi\sigma$, where σ is surface density of the electric charges serving as sinks and sources for the lines of the electric field. Hence, in magnetostatics I_n at the body boundary carries the same physical meaning as σ in electrostatics and therefore the lines of the field intensity $\mathbf{H}$ at the boundary undergo a break, exactly as it takes place in the theory of the magnetic charges [3, 10].

Equations (1.7), (1.10), (1.14) with boundary conditions (1.16-1.19) and definition (1.9) represent a complete system of equations allowing to calculate all magnetic properties of any magnetizable specimen. However, practical use of this approach to specimens of an arbitrary shape is restricted by complexity of the bound current distribution caused by inhomogeneous magnetization in such specimens. Fortunately, there is a large group of practically important specimens allowing rigorous analytical calculations of the magnetic properties. Those are the specimens of ellipsoidal shape. For this reason, unless otherwise indicated, in this book under the shapes of the specimens we will imply the shapes of ellipsoidal bodies, i.e., the bodies bound by an ellipsoidal surface.

As it was shown for the first time by Poisson [10, 14], in a body bound by a complete surface of a second degree (i.e., in ellipsoid) subject to a uniform field $\mathbf{H}_0$, the field intensity $\mathbf{H}$ is also uniform although it can be different from $\mathbf{H}_0$ (both in magnitude and direction). We will refer to this condition as Poisson theorem. A similar condition is valid for dielectric ellipsoids polarized by the applied electric field [7].

We remind that an ellipsoid is characterized by three mutually perpendicular axes a, b and c, and its volume is $V = 4\pi abc/3$. The difference between the applied field $\mathbf{H}_0$ and the field $\mathbf{H}$ inside the ellipsoidal specimen is referred to as demagnetizing field $\mathbf{H}_d$. This field is proportional to magnetization $\mathbf{I}$ and depends on the specimen geometry, i.e., on the relationship between the ellipsoidal axes. Introducing an orthogonal coordinate system with axes (x,y,z) directed along the axes (a,b,c)[5], respectively, the x-component of $\mathbf{H}_d$ in cgs units is

$$(H_d)_x = (H_0)_x - H_x = 4\pi I_x \eta_x, \qquad (1.20)$$

---

[5]Equation of the ellipsoidal surface in these coordinates is $(x/a)^2 + (y/b)^2 + (z/c)^2 = 1$.

where η_x is called *demagnetizing factor* with respect to that (x-) axis; η_x is a coefficient of proportionality between $(H_d)_x$ and I_x, and 4π is introduced for consistency with other formulae in *cgs* unit system. (In SI system 4π in Eq. (1.20) is omitted.) Replacing subscript x by y and z, the relationship given in Eq. (1.20) is valid for the y- and z-component, respectively.

The demagnetizing factors η_x, η_y and η_z are positive coefficients depending *only* on the shape of ellipsoid. One can show (see, e.g., [7, 10]) that

$$\eta_x + \eta_y + \eta_z = 1. \tag{1.21}$$

The simplest example of an ellipsoid is a sphere, for which $\eta_x = \eta_y = \eta_z = 1/3$ due to symmetry. Another example is a long circular cylinder, representing a strongly prolate ellipsoid of revolution. In the limit of $a \gg b = c$, the demagnetizing factor for the long axis is zero and that for two other axes (transverse to the long one) is $1/2$. If a base of the long cylinder is oval so that $a \gg b > c$, then $\eta_a \to 0$ and $\eta_b \to 1 - \eta_c < \eta_c$, where η_a, η_b and η_c are the demagnetizing factors with respect to coordinate axes parallel to corresponding axes of the ellipsoid. After all, if both a and b are much greater than c ($a \sim b \gg c$), then η_a and η_b are close to zero whereas $\eta_c \to 1$. In this limit, the base of the cylinder represents a very long and narrow rectangle and the entire ellipsoid represents an infinite parallel-plane plate (implying that the plate lateral dimensions greatly exceed its thickness). Apparently, the infinite plate can be also viewed as a strongly oblate ellipsoid with demagnetizing factor relative to the axis perpendicular to the plate $\eta_\perp = 1$ and the factors with respect to any axis parallel to the plate $\eta_\parallel = 0$. For illustration, cross-sectional views of specimens with different demagnetizing factors are shown in Fig. 1.1.

If the field $\mathbf{H}_0$ is parallel to one of the axes of the ellipsoidal specimen with respect to which the demagnetizing factor is η, then, as it is rigorously derived (using different

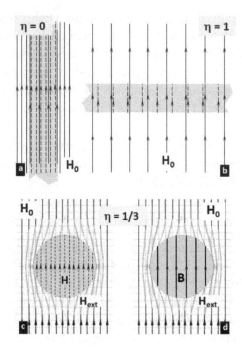

FIGURE 1.1 Cross-sectional view of ellipsoidal specimens (shown in gray) with different demagnetizing factors η in free space. (a) $\eta = 0$, a long cylinder in a parallel field or the specimen of a cylindrical geometry; (b) $\eta = 1$, an infinite plate in the perpendicular field or the specimen of a transverse geometry; (c) and (d) $\eta = 1/3$, a sphere. Lines outside the specimen are the lines of the outer $\mathbf{B}$ and $\mathbf{H}$ fields (in the free space $\mathbf{B} = \mathbf{H}$ or $\mathbf{B} = \mu_0\mathbf{H}$ in SI units); away from the specimen $\mathbf{B} = \mathbf{H} = \mathbf{H}_0$, the latter is an applied field; near the specimen the *external field* $\mathbf{H}_{ext}$ can differ from $\mathbf{H}_0$ as seen in (c) and (d). Inside the specimens, dashed and solid lines are the lines of the field intensity $\mathbf{H}$ and of the induction $\mathbf{B}$, respectively. Density of the $\mathbf{B}$ and $\mathbf{H}$ lines approximately corresponds to diamagnetic specimens with $\mu \approx 0.8$ in (a) and (b), and with $\mu \approx 0.5$ in (c) and (d). In all cases, $\mathbf{B}$ and $\mathbf{H}$ fields inside the specimens are uniform. In (a) $B < H = H_0$; in (b) $B = H_0 < H$; and in (c) and (d) $B < H_0 < H$.

approaches) in [10] and in [7], H and B inside and H_0 outside are parallel and related as

$$(1 - \eta)H + \eta B = H_0. \tag{1.22}$$

$$(1 - \eta)H + \eta \frac{B}{\mu_0} = H_0. \qquad \text{(SI)}$$

Heuristically, Eq. (1.22) can be derived as follows.

From (i) definition (1.9), (ii) continuity of the H_t at the specimen boundary (1.17), and (iii) condition $B = H = H_0$ away from a specimen, it follows that H, B and H_0 are linearly related. Therefore, when $\mathbf{H_0}$ is parallel to one of the ellipsoidal axes, then

$$w_1 H + w_2 B = H_0, \tag{1.23}$$

where w_1 and w_2 are coefficients depending on demagnetizing factor with respect to that axis η and *not* depending on the specimen material, implying that (1.23) holds for ellipsoidal specimens of *any* material.

Imagine that the specimen is a superconductor at temperature T above the critical temperature T_c. So, it is in the normal state implying that its magnetic permeability $\mu = 1$ and therefore everywhere $H = B = H_0$. Hence,

$$w_1 + w_2 = 1. \tag{1.24}$$

Now, let us cool this specimen down to $T < T_c$ and assume that H_0 is weak enough so that the specimen is in the Meissner state, i.e., $B = 0$ inside.

The magnetic moment, calculated from definitions (1.2) for I and (1.9) for H, is

$$M = VI = -\frac{V}{4\pi}H, \tag{1.25}$$

$$M = VI = -V \cdot H, \qquad \text{(SI)}$$

where the negative sign means that the moment is directed opposite to **H** (the field intensity inside the specimen).

On the other hand, from (1.20) we write

$$H = H_0 - H_d = H_0 - 4\pi I \eta. \tag{1.26}$$

Therefore,

$$M = -\frac{V}{4\pi\eta}(H - H_0). \tag{1.27}$$

Hence,

$$-\frac{V}{4\pi\eta}(H - H_0) = -\frac{V}{4\pi}H \tag{1.28}$$

and

$$H = \frac{H_0}{1 - \eta}. \tag{1.29}$$

Plugging (1.29) into (1.23), we obtain

$$w_1\frac{H_0}{1 - \eta} = H_0. \tag{1.30}$$

Therefore, $w_1 = 1 - \eta$ and $w_2 = 1 - w_1 = \eta$. Q.E.D.

If either B or H or relationship between them is known, Eq. (1.22) along with definition (1.9) allows one to calculate all magnetic properties of uniform ellipsoidal specimens (Problem 1.10). In particular, the magnetic moment of a specimen made of material with given μ is

$$M = -\frac{V}{4\pi}\frac{1 - \mu}{1 - \eta(1 - \mu)}H_0. \tag{1.31}$$

We see that magnetic properties (i.e., **M**, **B** and **H**) depend on the specimen material (μ) *and* on η, i.e., on the specimen shape and orientation of the applied field. For instance, the magnetic moment of an infinite slab in parallel field ($\eta = \eta_\parallel = 0$) is

$$M = -\frac{V}{4\pi}(1 - \mu)H_0. \tag{1.32}$$

Whereas the moment of the same slab in the perpendicular field $(\eta = \eta_\perp = 1)$ is

$$M = -\frac{V}{4\pi}\frac{(1-\mu)}{\mu}H_0. \tag{1.33}$$

In SI units 4π in Eqs. (1.31-1.33) should be dropped.

From (1.31) and also from (1.32) and (1.33) we see that M of paramagnetic bodies is always positive, whereas it is always negative for diamagnetics. Naturally, for dia- and paramagnetics, in which μ only barely differs from unity, there is almost no difference in magnitude of M measured in field of different orientations. However, magnetic properties of superconducting bodies, in which μ can be much less than unity (down to zero in the Meissner state), as well as properties of bodies made of ferromagnetic materials $(\mu \gg 1)$ very strongly depend on the body geometry and the field orientation.

At an arbitrary orientation of $\mathbf{H_0}$, it should be broken for components parallel to the ellipsoidal axes and the components of $\mathbf{M}$, $\mathbf{B}$ and $\mathbf{H}$ are calculated independently.

It is interesting to note that according to (1.33) M of the superconducting slab (assuming $\mu = 0$) in perpendicular field is infinite, and therefore magnetic energy of such a specimen $(E_m = -\mathbf{M}\cdot\mathbf{H}/2)$ is infinite as well, in apparent contradiction with the law of energy conservation. On that ground, Maxwell concluded that materials with $\chi = -1/4\pi$ (in SI units $\chi^* = -1$), i.e., with $\mu = 0$, do not exist [10]. In reality, however, this means that superconducting slabs in the perpendicular field can never be in the Meissner state, in full consistency with experiment (see, e.g., [23]). Pure superconducting plates in perpendicular field can be found only in an inhomogeneous state (i.e., in the intermediate state for type-I and the mixed state for type-II superconductors, respectively) regardless how small this field is.

Coming back to the non-conducting magnetizing materials, we summarize that (i) their magnetic properties can be described using the vector potential for $\mathbf{B}$ and the scalar potential for $\mathbf{H}$ (Problem 1.9); (ii) in specimens of ellipsoidal shapes the field intensity H is always homogeneous; (iii) the notions of the demagnetizing factor and the demagnetizing field are well defined for *uniform ellipsoidal* specimens (i.e., with $B = const$ throughout its interior), such as those of uniform dia- and paramagnetics, and superconductors in the Meissner state; (iv) for such specimens Eq. (1.22) holds regardless of the relationship between $\mathbf{B}$ and $\mathbf{H}$, which automatically includes cases when μ changes with the field.

1.2 CONDUCTORS

Conducting materials differ from dielectrics in ability to conduct a transport current, also referred to as conduction and free current. Hence, the mean density of microscopic currents in conductors can be presented as

$$\mathbf{j} = \mathbf{j}_c + \mathbf{j}_p, \tag{1.34}$$

where $\mathbf{j}_c$ is the mean density of the microscopic conduction current, which is the same as the regular transport current density used in a macroscopic theory. And $\mathbf{j}_p$ is the mean density of persistent (non-dissipating) microscopic currents circulating inside the body. It includes the bound currents and the currents induced in conduction electrons, leading to Pauli paramagnetism and Landau diamagnetism [21].

Pauli paramagnetism is associated with partial redistribution of the conduction electrons in the momentum space due to interaction of their spin magnetic moment with the magnetic field tending to align the spin moments with the field. However, due to the Pauli exclusion principle, only electrons with energy close to the Fermi energy can take part in this redistribution. Respectively, contribution of the Pauli paramagnetism in the total magnetic susceptibility of a metal is much less than that due to the bound electrons [16].

Landau diamagnetism is associated with the quantized orbital motion of the conduction electrons induced by the field (due to the Lorentz force). This leads to a number of remarkable macroscopic quantum phenomena taking place in very pure metals at low temperatures and high fields such as the Shubnikov-de Haas and the de Haas-van Alphen effects [24] and the quantum Hall effect in 2D specimens [25]. In weak fields, magnitude of the magnetic susceptibility due to Landau diamagnetism equals to one-third of that due to Pauli paramagnetism [26].

Specimens with transport current cannot be in thermodynamic equilibrium by definition. Therefore, we will consider only the cases when $\mathbf{j}_c$ is absent. On the other hand, conduction electrons just slightly vary the magnetization an therefore Eq. (1.3) remains in force provided $\mathbf{j}_b$ is replaced by $\mathbf{j}_p$. As a result, all formulas discussed in the previous section are valid for conducting magnetizing materials in the absence of a transport current.

1.3 SUPERCONDUCTORS

1.3.1 Meissner state

Superconductivity was discovered in the laboratory of Kamerlingh Onnes in 1911 [27] as phenomenon of disappearance of electrical resistivity in some metals (the first metal was mercury) at temperatures below so-called critical temperature T_c, a characteristic parameter of the superconducting material. In other words, below T_c and, as it turned out, below a temperature-dependent critical field[6] H_{cr}, a normally conducting metal with finite resistivity converts into super or perfect conductor with no resistivity whatsoever.

Soon after the discovery, this experimental fact was confirmed with several other metals, such as lead, tin and thallium. Since then, the number of superconducting materials is constantly growing (in the mid 1960s it was close to 1000 [28]), so at present it is hardly possible to list all of them.

---

[6]We denote the critical field as H_{cr} instead of commonly used H_c because the latter is reserved for a thermodynamic critical field, which can be different from H_{cr}.

This clearly indicates that superconductivity is a general and widespread phenomenon which mechanism(s) is(are) yet to be understood [29, 30, 31].

In the early years, in spite of complete oddity of the new phenomenon, a general consensus was that magnetic properties of superconductors are well understood. Supported by demonstration of a persistent current in a closed superconducting circuit [32], this consensus was based on the Faraday law of electromagnetic induction. This physics law (one cannot but agree with R. Feynman who considers its discovery to be the most significant event in the history of the nineteen century [22]) dictates that magnetic flux in a perfect conductor *must* be frozen-in. This means that if a specimen (for simplicity, assume a long cylinder) is cooled below T_c in zero field and after that the field $\mathbf{H}_0$ (assume parallel) is turned on, the electromotive force (e.m.f.) induced will keep the flux and therefore the induction inside the cylinder zero as it was before the field is applied. It is easy to show that the magnetic moment of this specimen will be $-(V/4\pi)H_0{}^7$. This implies that our perfectly conducting specimen behaves like a perfect diamagnetic with $\chi = -1/4\pi$ or $\mu = 0$.

However, if the same specimen is cooled down in non-zero field, then no e.m.f. is induced and therefore the flux and induction inside the specimen will stay unchanged (as they were before cooling). Respectively, magnetic moment in this case will be zero. If now the field is changed for some ΔH_0 (within the field range of the pure superconducting state), it will induce the e.m.f., which sets the current preventing the flux change, and the moment will be $M = -(V/4\pi)\Delta H_0$. Hence, M will be negative, like in diamagnetics, at the increasing field ($\Delta H_0 > 0$) or positive, like in paramagnetics, at the decreasing field ($\Delta H_0 < 0$).

---

[7]Strictly speaking, even in a perfectly conducting specimen (i.e., the specimen with an infinite electron mean free path) the magnetic moment should die out within a few hours due to scattering of electrons at the specimen surface [33].

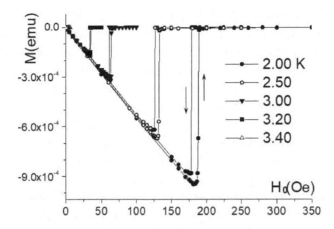

FIGURE 1.2 Experimental data for magnetic moment of a pure indium film 2.79 μm thick measured in parallel applied field H_0 at indicated temperatures [58]. Arrows indicate either the measurements were conducted at increasing (arrow up) or decreasing (arrow down) field. Transitions at $H_c(T)$ are first-order thermodynamic phase transitions between the pure superconducting (S) Meissner state and the normal (N) state. A hysteresis at the S/N phase transition is due to supercooling of the N phase.

Therefore, magnetic properties of a perfect conductor depend on history and therefore such a conductor can never be in thermodynamic equilibrium in a magnetic field. This picture was so obvious that it looked unreasonable investing resources for its experimental verification [35, 36].

This status of superconductivity lasted until 1932 when Keesom and co-workers succeeded in measurements of a specific heat of tin and thallium, where they revealed a sharp discontinuity at the critical temperature [37]. The jump in specific heat indicates that superconductivity can be associated with a yet unknown phase of matter and, therefore, the magnetic properties of superconductors can differ from those of perfect conductors, as it has been suggested on a number of occasions (see, e.g., [38]). The results of Keesom et al.

served as a powerful call to measure magnetic properties and corresponding experiments were set up in several groups.

First publications came from the groups of Walther Meissner [39] and Lev Shubnikov [40]. After the paper of Rjabinin and Shubnikov (see, e.g., [41] for details), it became clear that superconductivity is indeed a new thermodynamic phase in which zero resistivity is just a corollary of its magnetic properties [42]. Below we will abbreviate it as S phase and state. Magnetization curves $M(H)$ of genuine (i.e., pinning-free) superconductors are fully reversible, as must be the case for a thermodynamic equilibrium state in striking contrast with properties of a perfect conductor. In a weak magnetic field induction inside superconducting specimens is zero regardless on the field/cooling history. This is the essence of the Meissner (or Meissner-Ochsenfeld) effect. An example of magnetization curves for a specimen in the Meissner state is shown in Fig. 1.2[8]. The S state with $B = 0$ is referred to as the Meissner or pure superconducting state. Hence, superconductors in the Meissner state represent perfect diamagnetics.

However, the Meissner state is not the only equilibrium state of superconducting materials. The other states are briefly considered below in this chapter; in Ch. 3 the thermodynamic properties of specimens in these states will be discussed in more detail. It is important to note that magnetic moment of superconducting specimens in *any equilibrium state* is negative, i.e., at equilibrium any superconductor represents a diamagnetic.

A qualitative difference of the macroscopic properties of superconductors and "normal" diamagnetics is the presence of the critical field H_{cr} in the former, implying that superconductors transform into so-called normal (N) state at sufficiently high applied field. The N state or phase of superconducting materials is defined as the state with $\mu = 1$, which means that small diamagnetism or paramagnetism of materials in the N state is neglected.

---

[8]See section 1.3.3 for the nature of hysteresis at the S/N transition.

The S/N phase transition is a direct consequence of a finite magnitude of *condensation energy* E_c, the key thermodynamic property of superconductors defined as

$$E_c = F_{n0} - F_{s0}, \qquad (1.35)$$

where F_{n0} and F_{s0} are the specimen Helmholtz free energies of the N and S states in zero field, respectively, taken at the same temperature.

Note, since E_c is defined through the free energy at zero field, it does not matter through which thermodynamic potential (we discuss them in succeeding chapters) E_c is defined, provided the specimen compressibility is neglected (volume is assumed constant).

E_c is the energy gain received by the specimen when it transforms to the S state. It can be viewed as a "gold stock" allowing superconductors to "pay off" all their extraordinary properties, such as, e.g., the Meissner effect.

The physics behind E_c is associated with an energy gap in the spectrum of conduction electrons of a superconducting material. This and other microscopic properties of superconductors follow from the pairing of conduction electrons (creation of Cooper pairs) and are considered in the BCS (Bardeen, Cooper and Schrieffer) [43, 44] and Bogolubov theories [45, 46].

E_c is an additive (extensive) quantity determined by the material and depending on temperature only. The density of the condensation energy e_c is characterized by a *thermodynamic critical field* H_c defined as

$$e_c \equiv \left(\frac{\partial E_c}{\partial V}\right)_T = \frac{E_c}{V} = \frac{H_c^2}{8\pi} \qquad (1.36)$$

$$e_c \equiv \left(\frac{\partial E_c}{\partial V}\right)_T = \frac{E_c}{V} = \frac{B_c^2}{2\mu_0}. \qquad \text{(SI)}$$

The independence of E_c on the shape of the specimen leads to the fact that for any single-connected superconducting specimen in the magnetic field of any orientation

$$- \int_0^{H_{cr}} \mathbf{M} \cdot d\mathbf{H}_0 = \frac{H_c^2}{8\pi} V \qquad (1.37)$$

$$- \int_0^{B_{cr}} \mathbf{M} \cdot d\mathbf{B}_0 = \frac{B_c^2}{2\mu_0} V. \qquad (\text{SI})$$

This very important formula will be discussed in Ch. 3.

One should bear in mind, that the Meissner condition $B = 0$ is inapplicable to a surface layer over which B decays down to zero from its external value. Normally, the thickness of this layer (called "the magnetic penetration depth" λ) is of the order of 10-100 nm at $T \ll T_c$. So, the Meissner effect is observed in sufficiently massive specimens for which the volume of the penetration layer can be neglected. For this reason, the Meissner state does not exist in thin films. Also, as it was noted above, the Meissner effect does not take place in infinite slabs in perpendicular field schematically shown in Fig. 1.1b. Below we will refer to such specimen-field configuration as specimens of a *transverse geometry*. As it was indicated, the absence of the Meissner effect in this case follows from the law of energy conservation. The same conclusion can be easily retrieved from the law of flux conservation [47].

One should also underscore that superconducting material may behave as *superconductor*, in the sense of the thermodynamic phase of matter, only in the case of single-connected specimens like, e.g., solid cylinder, sphere, slab, *etc.* In multiple-connected specimens [3], such as, for instance, a ring, the flux passing through is trapped and these specimens behave as a perfect conductor[9]. Moreover, even in single-connected specimens, the flux can be trapped or pinned at so-called pinning centers, as it was for the first

---

[9]Detailed consideration of magnetic properties of a superconducting ring is available in [35].

time demonstrated in [40]. The pinning centers are different kinds of structural and chemical defects, such as vacancies, impurities and inhomogeneities in chemical composition. The pinning centers effectively work like openings in multiple-connected specimens. Overall, magnetic properties of multiple-connected and insufficiently pure single-connected specimens of superconducting materials represent properties of perfect conductors and have little in common with properties of the superconducting state of matter. In particular, thermodynamics is inapplicable to such specimens. This kind of "duplicity" of superconductors is a source of frequent confusion in interpretations of experimental results. Using Maxwell's wording, properties of superconductors must be "carefully distinguished" from properties of perfect conductors.

1.3.2 Interphase surface energy

In the Meissner state, the induction $B(= 0)$ is homogeneous. Contrarily, in other superconducting states, B is inhomogeneous: the specimen consists of domains of S and N phases. In the former $B = 0$ ($\mu = 0$) and in the latter $B = H$ ($\mu = 1$). Interfaces between domains, or more specifically the energy associated with the S/N interphase boundary, plays a crucial part in properties of superconductors. This energy, also called S/N surface tension, is calculated in a theory of Ginzburg and Landau (GL) [48]. The essence of this energy can be explained as follows [49].

Contacting S and N phases are bridged through some transition layer of diffuse nature, which thickness is referred to as domain-wall energy parameter δ. If this layer is on the S side from a physical interphase boundary (the boundary at which B vanishes), then superconductivity within it is partially suppressed and therefore part of condensation energy associated with this layer is lost. The domain-wall parameter δ and the interphase surface energy in this case are positive.

On the other hand, if the transition layer lays on the N side from the boundary $B = 0$, then it implies that the S phase partially penetrates into the N phase which makes δ and the S/N interphase energy negative.

The outlined scenarios are depicted in Fig. 1.3, where λ is the magnetic penetration depth and ξ is a Ginzburg-Landau coherence length characterizing a spatial extend of variation of number density of so-called superconducting electrons n_s. The latter is used in a theory of F. and H. London [36, 50] and corresponds to a fraction of "condensed" electrons in the two-fluid model of Gorter and Casimir [35, 51]. In the GL theory $n_s = \psi^2$, where ψ is a complex order parameter, which is the main quantity calculated in this theory. It should be noted that n_s is a derived concept, like number density of free electrons in theory of metals and fraction of superfluid component in He-II, and "does not represent the density of anything, which has microscopic meaning" [52].

Superconductors with positive and negative interphase surface energy are referred to as type-I and type-II superconductors, respectively. As one can see from Fig. 1.3, $\delta \sim (\xi - \lambda)$ and in type-I superconductors $\lambda < \xi$, whereas in type-II materials $\lambda > \xi$. The GL theory introduces an important parameter, allowing to distinguish different superconducting materials. This is the Ginzburg-Landau parameter $\kappa \equiv \lambda/\xi$. According to this theory, in type-I superconductors $\kappa < 1/\sqrt{2}$ and $\kappa > 1/\sqrt{2}$ in type-II materials.

The GL theory is discussed in detail in textbooks on superconductivity (e.g., in [12, 53, 54]). Here we just remind that the range of applicability of this theory is limited to a narrow vicinity of T_c by definition [48, 55] and thermodynamic potential (sometime referred to as Ginzburg-Landau free energy) is appropriate only for specimens of cylindrical geometry [56].

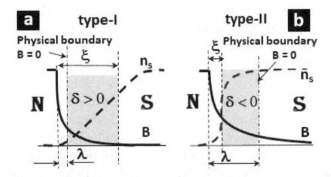

FIGURE 1.3 Schematic diagrams of variation of induction B and density of superconducting electrons n_s near the boundary between S and N phases in type-I (a) and type-II (b) superconductors. Designations: λ is the penetration depth, ξ is the coherence length and $\delta \sim (\xi - \lambda)$ is the domain wall parameter. Capital N and S designate pure normal and superconducting phases, respectively.

1.3.3 Type-I superconductors: Intermediate state

Type-I superconductors (materials with positive S/N surface tension) represent a relatively small group of superconductors mostly consisting of high purity elemental metals.

Since the S/N transition is associated with nucleation of a new phase (for example, S-phase) within an old one (N-phase in this example) and hence with creation of the S/N interphase boundaries, the positive S/N surface tension unambiguously leads to the fact that the S/N transition in type-I superconductors is discontinuous with the possibility of a metastable state of the old phase (in this example, possible "supercooling" of the N phase). This implies that the S/N transition in these materials is a thermodynamic phase transition of the first order, like, e.g., liquid-gas transition in ordinary matter.

For the same reason, in specimens of cylindrical geometry the positive S/N surface tension makes it thermodynamically unfavorable to create a two-component system. Hence, specimens of cylindrical geometry may have only one critical field, which is the critical field of the transition between the Meissner and the N states, implying that *for this geometry* $H_{cr} = H_c$.

However, as we already know, the Meissner state is not the only equilibrium state of a type-I superconductor. After Peierls [57], another state is referred to as intermediate state. The specimen in the IS is split into domains of S and N phases. Thermodynamic properties of the IS are discussed in Ch. 3; a recent review of experimental and theoretical studies of the IS is available in [58]. Here we outline the main characteristics of this state.

Imagine a long circular cylindrical specimen in a perpendicular field H_0. In a weak field, it is in the Meissner state, i.e., B over the specimen volume is zero and hence uniform. Therefore (while the specimen is in the Meissner state), its demagnetizing factor is well defined and equals $1/2$. The field lines inside and outside the specimen look similar to those shown in Figs. 1.1c and 1.1d except that there are no induction lines inside the cylinder in Fig. 1.1d and therefore (due to boundary condition for B_n (1.16)) the outer lines bend around the specimen. The outer field near the specimen surface, referred to as an external field H_{ext}, is (Problem 1.2)

$$H_{ext} = H_0 \frac{\sin \theta}{(1 - \eta)} = 2H_0 \sin \theta, \qquad (1.38)$$

where θ is the angle between the normal to the surface and $\mathbf{H}_0$.

As follows from the Poisson theorem (Eq. (1.22)), inside our cylinder H is uniform and equals to $H_0/(1 - \eta) = 2H_0$. Therefore, inside this specimen $H = H_c$ when the applied field is only $H_c/2$. Does this mean that in the perpendicular field, superconductivity in the cylinder is destroyed when the applied field is just a half of its value needed to destroy superconductivity in the parallel field? The answer is no, even if only because change of induction inside the specimen

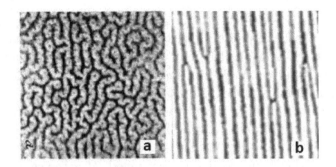

FIGURE 1.4 Examples of equilibrium domain patterns of the IS. (a) In an aluminum slab of 4.7 mm thickness in perpendicular field [61]; (b) in tin disc of 2 mm thickness in field tilted by 15° with respect to the specimen surface [62]. Scales are different. Dark areas are superconducting. (Reprinted with permission from The Royal Society (London).)

decreases induction outside it due to the flux conservation. This implies that neither discontinuous, nor gradual change of the induction at which the specimen core remains superconducting is possible (see, e.g., [35, 58] for more details).

The correct solution was for the first time suggested by Gorter and Casimir: at $H_0 > (1 - \eta)H_c$ the superconducting specimen is split into domains of S and N phases [59]. In our cylinder, the domains are discs parallel to H_0 [60]. In the general case equilibrium domains' shapes can be very diverse, from well ordered, as those in the cylinders in perpendicular field, to completely disordered as, e.g., in slabs in the perpendicular field. A common feature is that the domain walls in the specimen bulk always represent a surface formed by a straight generating line parallel to the field $\mathbf{H}$ and, as a consequence, to the induction $\mathbf{B}$ in the N domains. Examples of equilibrium domain patterns are shown in Fig. 1.4. Note that due to the boundary condition (1.17), the field intensity H is uniform throughout the bulk, despite the non-uniform induction B.

The first successful theoretical model of the IS was independently developed by Peierls [57] and F. London [36, 63]. The Peierls-London (PL) model (we will discuss it in Ch. 3) represents a global description of the IS in zero order approximation. The principle assumption in this model is that the field H in the ellipsoidal specimen equals to H_c over the entire field range of the IS (according to this model at $H_c(1-\eta) < H_0 < H_c$). Hence in the N domains $B = H = H_c$. On the other hand, as it was mentioned, H is the same both in the N and in S domains due to the boundary condition (1.17). Therefore, measuring B in N domains, one can (a) verify the assumption $H = H_c$ in the Peierls-London model and (b) determine H inside the S phase. This is all the more interesting because in normal materials, as it was indicated above, H is an unmeasurable quantity.

The induction B in the bulk of a high-purity specimen in the IS was measured by Egorov et al. [64]. Experimental data obtained are shown in Fig. 1.5. As one can see, B is always greater than H_0 and $B(H_0)$ saturates at the level $B = H_c$ when H_0 distances from H_{cr}. Hence, experimental results confirm that $H = H_c$ at low H_0; however, it decreases near H_{cr}. This change of $B(H_0)$ and other puzzling features of the IS unexplained in the Peierls-London model are accounted for in an advanced thermodynamic model of the IS [23, 34]. This model is also discussed in Ch. 3.

1.3.4   Type-II superconductors: Mixed state

As we already know, type-II superconductors are materials in which the S/N surface tension is negative. Coming from that, it is easy to trace a logical chain leading to potential thermodynamic profitability of a domain structure with maximum possible number of the N domains embedded in the S matrix, i.e., with minimal possible flux passing through each N domain. The latter is the flux quantum $\Phi_0 = \pi \hbar c/e = 2.07 \times 10^{-7}\text{G cm}^2$ ($\pi \hbar/e = 2.07 \times 10^{-15}$ Wb in SI units), where $\hbar = h/2\pi$ and h is the Plank constant.

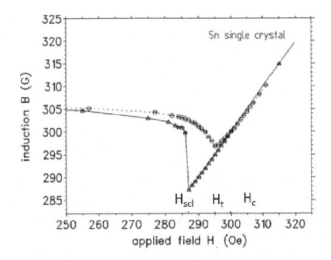

FIGURE 1.5 Induction in N domains of the Sn single crystal plate 0.56 mm thick at temperature 0.08 K measured at increasing (circles, solid line) and decreasing (triangles, dashed line) applied field. At decreasing field, the N state is field supercooled down to H_{scl}. At increasing field, the laminar structure transforms to one with tubular (filament) S regions at $H_t < H_c = 305$ G. At small applied field, B saturates at $B = H_c$. (After Egorov et al. [64], reprinted with permission from the American Physical Society.)

In type-I superconductors, in view of the positive S/N surface tension, the domain structure is not welcome (thermodynamically unfavorable) and therefore the N domains appear when the field H becomes equal H_c. In specimens of the cylindrical geometry, this leads to an immediate collapse of superconductivity over the whole volume since the stock of condensation energy is completely exhausted when $H = H_0 = H_c$. Contrarily, in type-II superconductors, due to negative S/N surface tension, the domains are welcome starting from some field H_{c1}. This field cannot be zero because energy of each domain, consisting of bulk and surface energy contributions, is positive but *less* than that in type-I superconductors. Therefore H_{c1} should be less than H_c and hence the domain structure can take place not only in specimens with $\eta \neq 0$, but also in the specimens of cylindrical geometry[10]. Then, from the mandatory condition Eq. (1.37) it follows that the magnetization curve of cylindrical specimens should have two critical fields: the low critical field $H_{c1} < H_c$ of a transition between the Meissner state and the inhomogeneous state with single-flux-quantum N domains; and the upper critical field $H_{c2} = H_{cr} > H_c$. Since the total flux Φ passing through the specimen changes by small steps $\Delta\Phi = \Phi_0$, one can expect that the S/N transition in type-II superconductors is continuous, i.e., a thermodynamic phase transition of second order.

This quantitative scenario is consistent with experimental data on the magnetization obtained in a fundamental study by Shubnikov[11] with co-workers [65], where type-II superconductivity was discovered, with the only exception that no one could imagine a negative surface tension at the time. Due to that, it took two decades before a coherent theoretical interpretation of these results was proposed by Abrikosov [56]. The concept of the S/N surface tension was introduced by H. London [66]; the idea of the negative S/N surface tension was for the first time explicitly addressed by Pippard [49].

After Abrikosov [56] superconductors which specimens of cylindrical geometry have two critical fields are referred to as superconductors of the second group or type-II superconductors. Using the GL theory, Abrikosov found that between H_{c1} and H_{c2} the flux structure represents a 2D-periodical lattice of single-flux-quantum N domains. The latter are referred to

---

[10]In case of zero surface tension, it would be thermodynamically indifferent whether the specimen of cylindrical geometry has or does not have domains, as well as how big these domains are.

[11]In 1937 36-year-old Lev Shubnikov, discoverer of the Shubnikov-de Haas effect, co-discoverer of the Meissner effect, discoverer of the type-II superconductivity and the author of a number of other fundamental discoveries, fell victim to the executioners of the soviet secret police; his last paper (L. Shubnikov and I. Nakhutin, Nature 139, 589 (1937)) was devoted to the problem of the intermediate state.

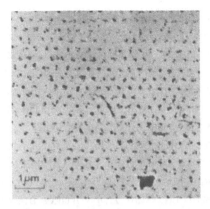

FIGURE 1.6 Equilibrium structure of the flux lines in a type-II superconductor observed on the end surface of a well annealed rod of Pb-4%In alloy at 1.1 K in a parallel field 3000 Oe. The black dots consist of small cobalt particles attracted to the regions with non-zero induction. (After Essmann and Träuble [67], reprinted with permission from Elsevier.)

as *vortices* and the S state at $H_{c1} < H < H_{c2}$ is referred to as the mixed state, the vortex state and the Shubnikov state.

The occurrence of the periodical vortex structure can be understood as follows. Consider, for simplicity, a specimen of transverse geometry. In such specimen, the MS starts immediately from $H_0 > 0$ no matter how small this field is, and there always will be plenty of flux lines (vortices). For instance, in a field $H_0 = 0.5$ G (the Earth field) the vortex number density $n_v = \overline{B}/\Phi_0 = H_0/\Phi_0 \sim 10^5$ mm^{-2} ($\overline{B}$ here is an average magnetic flux density). Then *due to symmetry*, at equilibrium the vortices should be arranged in the hexagonal 2D lattice in view of its highest symmetry among possible two-dimensional periodic structures and because any non-periodical structure of the flux lines is asymmetric.

In Fig. 1.6 we reproduce the first experimental image of an equilibrium flux structure in a type-II superconductor obtained by Essmann and Träuble [67]. We see that indeed the flux passes through the specimen in the form of lines arranged in the hexagonal lattice and the flux per one line, as reported in [67], is close to Φ_0. The flux structures consisting of the single-flux-quantum vortices, referred to as a vortex matter, is a general property of type-II superconductors both in equilibrium and non-equilibrium states (see, e.g., [47, 68, 69]). The huge difference in magnetic properties of the specimens without and with pinning (i.e., in and out of thermodynamic equilibrium, respectively) in the micron-order scale reduces to the fact that in the former case the vortices are ordered into a hexagonal lattice, while in the latter case the lattice ordering is disturbed. Near H_{c2} vortices are densely packed, hence forming the hexagonal lattice by definition. For this reason, irreversible experimental data on M vs. H_0 can be quite reversible close to H_{c2} and the purer the specimen, the wider is the range of reversibility starting from H_{c2} [65].

The vast majority of superconductors are type-II materials and the vast majority of them are alloys and multi-component compounds, such as, e.g., the Y-Ba-Cu-O high-temperature superconductors. Due to that, it is a tremendous challenge to fabricate pinning-free specimens of these materials necessary for the study of their genuine superconducting properties. There are, however, two type-II materials, Nb and V, which due to their elemental composition are the most appropriate for fabrication of pinning-free specimens. Nowadays the purest type-II specimens are niobium ones.

Typical reversible magnetization curves for cylindrical specimens of type-II superconductors are shown in Fig. 1.7. The magnetization curve of specimens of transverse geometry along with a phenomenological model consistently addressing equilibrium magnetic properties of type-II superconductors in specimens of any (ellipsoidal) shape are discussed in Ch. 3.

1.3.5 Nucleation of superconductivity

Systematic studies of magnetic properties of type-II superconductors were started in 1960s (see, e.g., [70]). It was re-

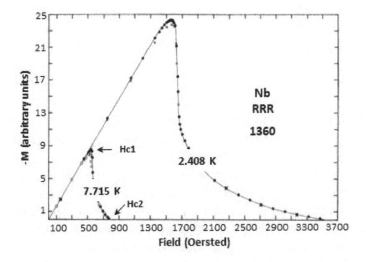

FIGURE 1.7 Reversible magnetization curves for a specimen of cylindrical geometry of a type-II superconductor (niobium). Solid circles are experimental points obtained at ascending field; open circles and crests are the points measured at descending field. RRR (residual resistivity ratio) indicates for an extremely high purity of the specimen. (After Finnemore at al. [70], reprinted with permission from the American Physical Society.)

vealed that H_{cr} determined from resistivity measurements differs quite significantly (about a factor of 2 and even greater) from that determined from magnetization data. Saint-James and De Gennes [71], based on calculations within the GL theory, proposed that this difference is due to influence of specimen surface on nucleation of superconductivity at high field.

According to Saint-James and De Gennes, in specimens of cylindrical geometry (i.e., in parallel field) superconductivity nucleates near the surface at $H_{c3} = 1.7H_{c2}$. However, in specimens of the transverse geometry (i.e., in perpendicular field) the nucleation occurs in the bulk at H_{c2}. Between

H_{c2} and H_{c3} the superconducting phase forms a thin continuous surface sheath. The latter is not visible in magnetization measurements due to limited sensitivity of magnetometers available at the time. However, this sheath short-circuits the potential leads at measurements of resistivity. Hence, in cylindrical specimens the S/N critical field determined through magnetization measurements is H_{c2}, whereas the critical field following from the resistance measurements is H_{c3}.

In spite of its general acceptance [11, 53, 54], this surface superconductivity interpretation has left quite a few questions. For instance, by definition, the field passes through specimens in the N state being unperturbed, i.e., not noticing the surface. Then, why does the nucleation of superconductivity depend on the field to surface orientation?

Five decades after the paper of Saint-James and De Gennes, the problem of superconductivity above H_{c2} was revisited by Kozhevnikov et al. [72] in a complex of experiments with use of high-purity niobium specimens in both parallel and transverse fields and of the state-of-the-art techniques for measuring magnetization, electrical resistance, induction and imaging of the near-surface flux distribution. It was shown that superconductivity nucleates in the specimen *bulk* at $H_{c3} \approx 2H_{c2}$ regardless of the field orientation; between H_{c2} and H_{c3}, the superconducting phase exists in the form of filaments parallel to the applied field. Increasing the field above H_{c2} the filament number density decreases vanishing at H_{c3}.

Traces of superconductivity above H_{cr} also take place in type-I superconductors [23, 64]. In particular, they are well seen in Fig. 1.5 at ascending field. At descending field, these traces are absent due to supercooling of the N state caused by the positive S/N surface tension.

The volume of the superconducting phase and the corresponding amount of the condensation energy associated with superconductivity above H_{cr} can be estimated from the area under magnetization curves reported in [72]. This yields about 1% of the total condensation energy (1.37).

Our overview of superconductivity would not be complete without mentioning experimental works in which superconductivity was discovered at pressures in the millions-atmosphere range (see [73] for review). In particular, the record high $T_c (\approx 250\,\text{K})$ was reportedly observed in lanthanum hydride LaH_{10} at $P \approx 170\,\text{GPa}$ [74]. Unfortunately, only limited studies with use of tiny specimens are possible at such conditions. However, as of today, there is no reason to believe that superconducting properties at high pressure differ from those at normal conditions, apart from peculiar structural phases of some materials not yet found at ambient conditions.

1.4 PROBLEMS

1.1. A long cylindrical specimen with a base of arbitrary shape is subjected to a uniform magnetic field $\mathbf{H}_0$ parallel to its longitudinal axis. The cross-section of the specimen and the field lines inside and near it look as shown in Fig. 1a. The magnetic permeability of the specimen material is μ. The specimen is in vacuum. Find the force acting on unit area of its lateral side.

Solution. The force $\mathbf{f}$ acting on unit area of a surface between two contacting media subjected in electric and/or magnetic field(s) can be calculated using the Maxwell stress tensor originally derived from a model of an elastic aether [10]. Justification of this approach can be found in [3, 8]. In the magnetic field, this force is

$$\mathbf{f} = \frac{1}{4\pi}(B_{2n}\mathbf{H}_2 - B_{1n}\mathbf{H}_1) - \frac{1}{8\pi}(B_2 H_2 - B_1 H_1)\mathbf{n} \quad \text{(p1.1-1)}$$

$$\mathbf{f} = (B_{2n}\mathbf{H}_2 - B_{1n}\mathbf{H}_1) - \frac{1}{2}(B_2 H_2 - B_1 H_1)\mathbf{n}, \quad \text{(SI)}$$

where μ_1 and μ_2 are permeabilities of the contacting (1 and 2, respectively) media, $\mathbf{n}$ is unit vector directed outward with respect to the medium 1; $\mathbf{H}_1$ and $\mathbf{H}_2$ are the near-boundary field intensities in the first and second media and B_{1n} and B_{2n} are normal components of the fields $\mathbf{B}_1$ and $\mathbf{B}_2$ (near-boundary inductions in the first and second media, respectively).

The demagnetizing factor of the given specimen is $\eta = 0$. Then, outside the specimen $\mathbf{H}_2 = \mathbf{B}_2 = \mathbf{H}_0$ and inside it $\mathbf{H}_1 = \mathbf{H}_0$ and $\mathbf{B}_1 = \mu\mathbf{H}_1 = \mu\mathbf{H}_0$. All fields are parallel to the surface and therefore $B_{2n} = B_{1n} = 0$. Plugging these conditions in (p1.1-1), one obtains

$$\mathbf{f} = -\frac{H_0^2}{8\pi}(1 - \mu)\mathbf{n}$$

$$\mathbf{f} = -\frac{B_0^2}{2\mu_0}(1 - \mu)\mathbf{n}. \quad \text{(SI)}$$

Thus, a specimen of a diamagnetic material experiences compression (positive pressure) and that of a paramagnetic material is decompressed (experiences negative pressure). For superconductors of cylindrical geometry in the Meissner state, the outside pressure equals the energy density of the applied field.

1.2. A magnetic field $\mathbf{H}_0$ is parallel to one of the axes of a superconducting specimen with a demagnetizing factor η. The specimen is in the Meissner state. Using Eq. (1.22), find (a) the field intensity H inside the specimen; (b) the specimen magnetic moment M; (c) the outer field H_{ext} near the specimen surface.

Solution.

(a) Since $\mathbf{H}_0$ is parallel to the specimen axis, $\mathbf{H}$ is parallel to $\mathbf{H}_0$. From Eq. (1.22) and the Meissner condition $B = 0$ the magnitude of $\mathbf{H}$ is

$$H = \frac{H_0}{(1 - \eta)}.$$

Note, that η is well defined since $B(= 0)$ is uniform.

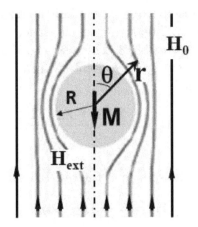

FIGURE 1.8 Superconducting sphere in the Meissner state in Problem 1.3. $\mathbf{H}_0$ is the applied field; $\mathbf{M}$ and $\mathbf{r}$ are the specimen magnetic moment and a radius vector in Eq. (p1.3-1), respectively; θ and R are the angle and the radius in Eq. (p1.3-2), respectively.

(b)

$$M = V\frac{(B - H)}{4\pi} = -V\frac{H_0}{4\pi(1 - \eta)}.$$

(c) Due to the boundary condition for B_n (1.16) and the Meissner condition $B = 0$, the outer field near the specimen H_{ext} is tangential to its surface, i.e., the field lines bend around the specimen. Tangential component of the inner field H is $H\sin\theta$, where θ is the angle between $\mathbf{H}_0$ and the normal to the surface. Hence, from the boundary condition for H_t (1.17), it follows

$$H_{ext} = H\sin\theta = \frac{H_0}{(1 - \eta)}\sin\theta.$$

1.3. (a) Using formula for the field of a magnetic dipole, calculate magnetic moment M of a spherical specimen in the Meissner state depicted in Fig. 1.8. (b) Show validity of

Eq. (1.22) for this specimen. (c) Find the magnetic field outside the specimen. The radius of the specimen is R and demagnetizing factor $\eta = 1/3$. The field $\mathbf{H}_M$ due to a magnetic
dipole with moment $\mathbf{M}$ at a point with radius-vector $\mathbf{r}$ from
the dipole center is

$$\mathbf{H}_M = \frac{3(\mathbf{M} \cdot \mathbf{r})\mathbf{r}}{r^5} - \frac{\mathbf{M}}{r^3}. \qquad (\text{p1.3-1})$$

Solution
(a) Outside the specimen the, field H' is

$$\mathbf{H}' = \mathbf{B}' = \mathbf{H}_0 + \mathbf{H}_M.$$

From the boundary condition (1.17) along with the Meissner condition $B = 0$ inside the specimen, it follows that the
normal component of B' at $r = R$ is zero. Hence, referring
to Fig. 1.8 we write

$$H_0 \cos\theta + \frac{3MR\cos\theta}{R^5}R - \frac{M\cos\theta}{R^3} = 0 \qquad (\text{p1.3-2})$$

So,

$$M = -\frac{R^3}{2}H_0 = -\frac{3}{2}\frac{V}{4\pi}H_0 = -\frac{V}{4\pi(1-1/3)}H_0,$$

where the negative sign means that M is anti-parallel to H_0.
On the other hand M, by definition, is

$$M = VI = V\frac{(B-H)}{4\pi} = -V\frac{H}{4\pi}.$$

Comparing the last two expressions, we see that

$$H_0 = (1 - 1/3)H = (1 - \eta)H.$$

This is identical to (1.22) for $B = 0$. Q.E.D.
(b) The field outside the specimen is

$$\mathbf{B}' = \mathbf{H}' = \mathbf{H}_0(1 + \frac{R^3}{2r^3}) - \frac{3R^3}{2r^5}(\mathbf{H}_0 \cdot \mathbf{r})\mathbf{r}$$

In SI units

$$\mathbf{B}' = \mu_0 \mathbf{H}' = \mathbf{B}_0 (1 + \frac{R^3}{2r^3}) - \frac{3R^3}{2r^5}(\mathbf{B}_0 \cdot \mathbf{r})\mathbf{r}, \quad (SI)$$

where $\mathbf{B}_0 = \mu_0 \mathbf{H}_0$ and $r \geq R$.

We leave it for the reader to show that $\mathbf{B}'(R)$ is the same as $\mathbf{H}_{ext}$ calculated in Problem 1.2.

1.4. A long cylindrical solenoid consists of a single layer of tightly wound turns of a thin wire. The solenoid length and radius are L and R, respectively; the wire diameter is D, and the number of turns is N. Assume $L \gg R \gg D = L/N$. The solenoid is in free space and a uniform magnetic field H_0 is applied parallel to its longitudinal axis.

(a) What current J_0 should be set in solenoid to zero the induction B inside it? (b) Find the magnitude of the linear density of this current g_c. (c) What is the field strength H inside the solenoid? (d) What are B and H outside the solenoid? (e) What is magnetic moment of a unit length (M/L) of the solenoid?

Solution.

(a) From either the boundary condition (1.17a) or Ampere's Law, in absence of the applied field, the field inside the solenoid $H_i = B_i$ is

$$B_i = \frac{4\pi}{c} n J_0,$$

where $n = N/L$.

When the parallel applied field H_0 is turned on the induction inside the solenoid becomes $B = B_0 + B_i = 0$, where $B_0 = H_0$. So the current strength J_0 needed to zero B is

$$J_0 = \frac{c}{4\pi} \frac{H_0}{n}.$$

(b)

$$g_c = \frac{J_0 N}{L} = J_0 n = \frac{c}{4\pi} H_0.$$

(c) Since the solenoid is in free space, inside it

$$H_i = B_i = 0.$$

(d) Since the solenoid is long (infinite), outside it

$$B = H = H_0.$$

Based on answers (c) and (d), one can say that this current screens the solenoid interior from the outer field.

(e)

$$\frac{M}{L} = -\frac{1}{L} \cdot N \frac{J_0 \pi R^2}{c} = -n J_0 \frac{\pi R^2}{c} = -\frac{1}{L}\frac{V}{4\pi}H_0,$$

where $V = \pi R^2 L$ is the solenoid volume. The sign $(-)$ means that the direction of J_0 is such that the field $\mathbf{H}_i$ and therefore the moment $\mathbf{M}$ are anti-parallel to $\mathbf{H}_0$.

1.5. A superconducting specimen has a shape of a right circular cylinder (a cylinder with a circular base perpendicular to the cylinder longitudinal axis) with the same dimensions (L and R) as those of the solenoid of the previous problem. The specimen is in vacuum and the field H_0 is applied parallel to its longitudinal axis. $H_0 < H_{c1}$, meaning that the specimen is in the Meissner state and its demagnetizing factor $\eta = 0$.

(a) Find the linear density of the surface current $\mathbf{g}$. (b) What are the magnitudes of B, H and M for this specimen? (c) Are the magnetic properties of the superconducting cylinder identical to those of the solenoid of the previous problem? If not, then which properties are the same and which ones are different?

Answer: (a) $\mathbf{g} = -c\mathbf{H}_0 \times \mathbf{n}/4\pi$; (b) inside $\mathbf{B} = 0$ and $\mathbf{H} = \mathbf{H}_0$, outside $\mathbf{B} = \mathbf{H} = \mathbf{H}_0$; $\mathbf{M} = -V\mathbf{H}_0/4\pi$. (c) No: the field intensity H inside the superconducting cylinder ($= H_0$) is different from that in solenoid ($= 0$). Values of the rest properties are quantitatively the same, but $\mathbf{g}_c$ in solenoid is linear density of a transport (dissipating) current supported by a battery, whereas g in the superconductor is linear density of a persistent (non-dissipating) current. Correspondingly, the boundary condition for H_t in the solenoid is Eq. (1.17a) and therefore there is the jump in H; whereas the boundary conduction for H_t in the superconductor is given by Eq. (1.17) and therefore H inside the specimen equals H_0.

1.6. Find the linear density of the surface current g in a spherical superconducting specimen in the applied field $H_0 < 2H_{c1}/3$ (the specimen is in the Meissner state).

Solution.

Coming from the boundary condition for the tangential component of the induction (1.18) and using H_{ext} calculated in problem 1.2, we write

$$g = \frac{c}{4\pi(1-\eta)}H_0\sin\theta = \frac{3c}{8\pi}H_0\sin\theta.$$

The direction of **g** is found from Eq. (1.4). It is clockwise if viewed from the tip of the vector **H**$_0$.

1.7. A non-magnetic ($\mu = 1$) ball of radius R is wrapped with a thin wire so that the turns are parallel to each other as schematically shown in Fig. 1.9. Assume that each turn is a closed and electrically isolated circular loop. Now imagine that there are tiny batteries in the loops so that the current in each of them is $J_\theta = g_c D$, where the wire diameter $D \ll R$ and g_c is the linear density of the surface transport current equal to g found in the previous problem. The currents runs clockwise if looking from the top pole (p. P). So, the loops make a shell with currents along the parallels. From the Poisson theorem we know that in a superconducting sphere with such a surface current the field H is uniform. Hence, one can expect that the field inside this current shell (where $H = B$) is uniform as well. The latter can be also seen from symmetry reasons. Verify this statement by calculating the field in the center of the current shell (p. O) and along the polar diameter PP'.

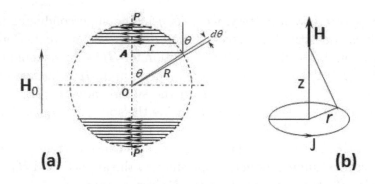

(a) **(b)**

FIGURE 1.9 (a) Schematics of the current shell for problems 1.7 and 1.8. Horizontal lines depict the circular loops (not all loops are shown) lying in horizontal (perpendicular to the page) plane with current flowing clockwise, as shown by arrows. H_0 is the applied field. (b) Schematic of a magnetic field H at a distance z from a center of a circular loop of radius r carrying a steady current J.

Hint: The field at a distance z above the center of a circular loop of radius r with a current J (Fig. 1.9b) is

$$H = \frac{2\pi r^2 J}{c(r^2 + z^2)^{3/2}}$$

$$H = \frac{r^2 J}{2(r^2 + z^2)^{3/2}} \quad \text{(SI)}$$

Solution.
A. Field at p. O.
Consider an arbitrary loop centered at p. A. The current dJ in this loop is

$$dJ = J_\theta = g_c R d\theta = \frac{3cRH_0}{8\pi} \sin\theta d\theta$$

So, the field dH_O due to this current at p. O is

$$dH_O = \frac{2\pi r^2 dJ}{cR^3} = \frac{2\pi (R^2 \sin^2\theta)dJ}{cR^3} = \frac{3H_o}{4} \sin^3\theta d\theta$$

And the total field H_O at p. O is

$$H_O = \frac{3H_o}{4} \int_0^\pi \sin^3 \theta d\theta = H_0.$$

B. Field at p. P.

Referring to the same Fig. 1.9a, we write for the field dH at p. P due to the current in the same single loop centered at p. A

$$dH = \frac{2\pi(R^2 \sin^2 \theta^2)dJ}{c(R^2 \sin^2 \theta + R^2(1 - \cos\theta)^2)^{3/2}} = \frac{3H_o}{4} \cos^3(\frac{\theta}{2})d\theta$$

Therefore, for the field H_P at p. P we obtain

$$H_P = \frac{3H_o}{4} \int_0^\pi \cos^3(\frac{\theta}{2})d\theta = \frac{3H_o}{2} \int_0^{\pi/2} \cos^3 \varphi d\varphi = H_0.$$

C. Field in p. P'.

Due to symmetry, the field in the lower pole P' is the same as that in the upper pole P. Hence, $H_{P'} = H_0$.

Taking into account the direction of the currents, we see that the field in the center and at the poles is the same and equals $-\mathbf{H}_0$. Therefore, the field along PP' is constant and equals $-\mathbf{H}_0$. We leave it to the reader to show that the field is the same $-\mathbf{H}_0$ in any other point inside the shell. As a hint, one can calculate the field at an inner equatorial point using the boundary condition (1.17a) and/or calculate the field in an arbitrary point inside the shell. In the latter case it is useful first to consider Problem 2 after §3 in [7] or Problem 5.48 in [8], 3d Edition.

1.8. (a) What is the magnetic moment of the current shell considered in the previous problem? (b) If the shell is placed in a vertically directed magnetic field H_0 (Fig. 1.9), what are the fields H and B inside and outside the shell?

Solution. (a) The magnitude of the magnetic moment dM due to the current in a single loop of the previous problem is

$$dM = \frac{\pi r^2 dJ}{c} = \frac{\pi R^2 \sin^2 \theta}{c} \cdot \frac{3cH_0 R}{8\pi} \sin\theta d\theta$$

Hence, the magnetic moment of the shell is

$$M = \frac{3H_0 R^3}{8} \int_0^\pi \sin^3 \theta d\theta = \frac{H_0 R^3}{2} = \frac{3}{2}\frac{V H_0}{4\pi} = \frac{V H_0}{4\pi(1 - 1/3)},$$

where $V = 4\pi R^3/3$ is volume inside the shell.

Note that M of the shell is the same as that of superconducting sphere ($\eta = 1/3$) in the Meissner state following from Eq. (1.32), as it should be.

(b) Taking into account the direction of the field produced by the current in the loops (downward in Fig. 1.9) and the fact that μ inside the shell is unity, the fields inside are

$$\mathbf{H} = \mathbf{H}_0 + \mathbf{H}_{shell} = 0$$

$$\mathbf{B} = \mu\mathbf{H} = 0$$

Since the shell magnetic moment is the same as that of the spherical superconductor in the field H_0 and the shell is also in the field H_0, the outer field of the shell is the same the field outside the superconductor. So the outer field near the shell is

$$B_{ext} = H_{ext} = \frac{H_0}{(1 - \eta)} \sin\theta = \frac{3H_0}{2} \sin\theta.$$

We leave it for the reader to check this answer using the formula for the field H_M of the magnetic moment given in Problem 1.3.

The field far away from the shell is

$$B = H = H_0.$$

1.9. Using magnetic (a) vector and (b) scalar potentials, calculate the magnetic field due to a system of closed currents with the magnetic moment $\mathbf{M}$ at a point with radius vector $\mathbf{r}$, which length r significantly exceeds the size of the system. The system is in vacuum.

Solution.

(a) The vector potential $\mathbf{A}$ in some point P due to a system of currents with density $\mathbf{j}$ circulating in an arbitrary volume V is (see, e.g., [3, 8])

$$\mathbf{A} = \frac{1}{c} \int_V \frac{\mathbf{j}dV}{r'},$$

$$\mathbf{A} = \frac{\mu_0}{4\pi} \int_V \frac{\mathbf{j}dV}{r'} \quad \text{(SI)},$$

where r' is the distance from a current element $\mathbf{j}dV$ to point P.

For the distances significantly exceeding the system size, the above expression for $\mathbf{A}$ is transformed to

$$\mathbf{A} = \frac{\mathbf{M} \times \mathbf{r}}{r^3},$$

$$\mathbf{A} = \frac{\mu_0}{4\pi} \frac{\mathbf{M} \times \mathbf{r}}{r^3} \quad \text{(SI)},$$

where $\mathbf{r}$ is the radius-vector of the point P with origin somewhere inside the system of currents characterized by the dipole magnetic moment $\mathbf{M}$.

The magnetic induction $B(= H$ since $\mu = 1)$ is calculated from the defining relationship

$$\mathbf{B} = \mathbf{H} = \nabla \times \mathbf{A} = \nabla \times (\frac{\mathbf{M} \times \mathbf{r}}{r^3}) =$$

$$\frac{1}{r^3}\nabla \times (\mathbf{M} \times \mathbf{r}) + \nabla(\frac{1}{r^3}) \times (\mathbf{M} \times \mathbf{R}) =$$

$$\frac{3(\mathbf{M} \times \mathbf{r})\mathbf{r}}{r^5} - \frac{3\mathbf{M}}{r^3}.$$

Here we used a general relationship

$$\nabla \times (a\mathbf{b}) = a(\nabla \times \mathbf{b}) + (\nabla a) \times \mathbf{b}$$

and

$$\nabla(\frac{1}{r^3}) = -\frac{3\mathbf{r}}{r^5}.$$

(b) Absence of currents outside the system allows one to calculate the field $H(= B)$ using scalar potential Ψ. This yields

$$H = B = -\nabla\left(\frac{\mathbf{M} \cdot \mathbf{r}}{r^3}\right) = \frac{3(\mathbf{M} \cdot \mathbf{R})\mathbf{r}}{r^5} - \frac{\mathbf{M}}{r^3}.$$

Results of both approaches are identical. Q.E.D.

1.10. An ellipsoidal specimen is placed in a uniform magnetic field $\mathbf{H}_0$ directed parallel to one of the specimen axes. The magnetic permeability of the specimen material is μ and the specimen demagnetizing factor is η. The specimen is in free space.

(a) What are the induction $\mathbf{B}$ and the field intensity $\mathbf{H}$ inside the specimen?

(b) What are the magnitudes of the induction B_{ext} and of the field intensity H_{ext} outside the specimen near its surface?

(c) What is the linear density of the surface current $\mathbf{g}$?

(d) What is the specimen magnetic moment?

Hint: Use Eq. (1.22) and the boundary conditions for H and B.

Solution.

(a) Plugging $B = \mu H$ (or $B = \mu\mu_0 H$ in SI units) into Eq. (1.22), we obtain

$$\mathbf{H} = \frac{\mathbf{H}_0}{1 - \eta(1 - \mu)}. \qquad (\text{p}\,1.10\text{-}1)$$

and

$$\mathbf{B} = \frac{\mu\mathbf{H}_0}{1 - \eta(1 - \mu)}. \qquad (\text{p}\,1.10\text{-}2)$$

$$\mathbf{B} = \frac{\mu\mathbf{B}_0}{1 - \eta(1 - \mu)}. \quad (\text{SI})$$

(b) Denoting the fields on the inner and outer sides of the specimen surface as ($\mathbf{H}$ and $\mathbf{B}$) and ($\mathbf{H}'$ and $\mathbf{B}'$), respectively, for the normal components of $\mathbf{B}$ we write

$$B'_n = B_n = \frac{\mu H_0 \cos\theta}{1 - \eta(1 - \mu)}$$

and for the tangential

$$B'_t = H'_t = H_t = \frac{H_0 \sin \theta}{1 - \eta(1 - \mu)}.$$

Hence, the magnitude of the external field near the specimen surface is

$$B_{ext} = H_{ext} = \frac{H_0}{1 - \eta(1 - \mu)} \sqrt{\mu^2 \cos^2 \theta + \sin^2 \theta} \quad \text{(p 1.10-3)}$$

$$B_{ext} = \mu_0 H_{ext} = \frac{\mu_0 H_0}{1 - \eta(1 - \mu)} \sqrt{\mu^2 \cos^2 \theta + \sin^2 \theta}. \quad \text{(SI)}$$

And the angle φ between H_{ext} and the normal to the surface is

$$\tan \varphi = \frac{B'_t}{B'_n} = \frac{\tan \theta}{\mu} \quad \text{(p 1.10-4)}$$

(c) Using Eq. (1.4) we obtain

$$\mathbf{g} = -\frac{c}{4\pi} \frac{(1 - \mu) \sin \theta}{1 - \eta(1 - \mu)} H_0 \times \mathbf{n}. \quad \text{(p 1.10-5)}$$

(d) M calculated from definition (9) is

$$\mathbf{M} = -\frac{V}{4\pi} \frac{(1 - \mu)}{1 - \eta(1 - \mu)} \mathbf{H}_0 \quad (1.10\text{-}6)$$

For superconductors in the Meissner state ($\mu = 0$), Eqs. (p1.10-1) - (p1.10-6) become

$$\mathbf{H} = \frac{\mathbf{H}_0}{(1 - \eta)}$$

$$\mathbf{B} = 0$$

$$B_{ext} = H_{ext} = \frac{H_0 \sin \theta}{(1 - \eta)}$$

$$\varphi = \pi/2$$

The external field is tangential to the specimen surface; in other words, it bends around the specimen.

$$\mathbf{g} = -\frac{c\sin\theta}{4\pi(1-\eta)}\mathbf{H}_0 \times \mathbf{n}$$

and

$$\mathbf{M} = -\frac{V\mathbf{H}_0}{4\pi(1-\eta)}$$

After all, for specimens of cylindrical geometry these quantities are

$$\mathbf{H} = \mathbf{H}_0,$$

$$\mathbf{B} = 0,$$

$$B_{ext} = H_{ext} = H_0,$$

$$\varphi = \pi/2,$$

$$\mathbf{g} = -\frac{c}{4\pi}\mathbf{H}_0 \times \mathbf{n},$$

$$\mathbf{M} = -\frac{V\mathbf{H}_0}{4\pi}.$$

We leave it for the reader to verify how these values change as the specimen material changes from diamagnetic to paramagnetic.

CHAPTER 2

Thermodynamics of Para- and Diamagnetics

2.1 WORK AND ENERGY IN MAGNETIC FIELD

Thermodynamics operates with *thermodynamic potentials* or *functions of* (parameters of) *state*, whose infinitesimal increment is a total differential of these parameters. In the state of thermodynamic equilibrium, the thermodynamic potential(s) of a system is(are) minimum with respect to its(their) values for other possible states of the system. In traditional thermodynamics (mostly theory of heat), the form of thermodynamic potentials can be derived from principles of statistical physics by Gibbs. However, it is not the case for thermodynamics of a system involving vector fields, where the form of the function of state depends on the shape of an object of interest and on the field orientation. This makes the choice and/or construction of the function of state appropriate for such systems a non-trivial part of calculation of equilibrium properties. In this chapter, we consider thermodynamic potentials appropriate for specimens of magnetizing non-superconducting materials in static magnetic fields of different orientation.

Application of the magnetic field changes energy of a system and therefore requires a work done by an external agency. In isothermal processes, the role of the latter is played by the magnet power supply. As mentioned in the previous chapter, work needed to change the induction for δB is

$$\delta W = \frac{1}{4\pi} \int \mathbf{H}\,\delta\mathbf{B}\,dV \qquad (2.1)$$

$$\delta W = \int \mathbf{H}\,\delta\mathbf{B}\,dV, \qquad \text{(SI)}.$$

where the integral is taken over the system volume, i.e., over the whole space occupied by the field, often extending to infinity.

Detailed discussion and justification of this formula is available in [3]. Its simplified derivation can be as follows [15].

Since magnetic field penetrates generating its sources, a system with magnetic field includes current in the magnet. Then, the Maxwell equations for curl of the field intensity **H** in absence of the static electric field, and for curl of the vortex electric field **E** induced due to changing induction are

$$c\nabla \times \mathbf{H} = 4\pi \mathbf{j}_c \qquad (2.2)$$

$$\nabla \times \mathbf{H} = \mathbf{j}_c \qquad \text{(SI)}$$

$$c\nabla \times \mathbf{E} = -\frac{\partial \mathbf{B}}{\partial t} \qquad (2.3)$$

$$\nabla \times \mathbf{E} = -\frac{\partial \mathbf{B}}{\partial t}, \qquad \text{(SI)}$$

where $\mathbf{j}_c$ is density of the transport current, which is current in the magnet, and t is time.

Multiply Eqs. (2.2) and (2.3) by $(\mathbf{E}/4\pi)dV$ and $(\mathbf{H}/4\pi)dV$, respectively, subtract (2.3) from (2.2) and integrate over the entire volume taken by the magnetic field. Assuming that the field vanishes at the field boundary[1] we obtain

$$\int \mathbf{j}_c \mathbf{E}\, dV + \frac{1}{4\pi} \int \frac{\partial \mathbf{B}}{\partial t} \mathbf{H}\, dV =$$

$$= \frac{c}{4\pi} \int [\mathbf{E} \cdot \nabla \times \mathbf{H} - \mathbf{H} \cdot \nabla \times \mathbf{E}]dV =$$

$$= \frac{c}{4\pi} \int \nabla \cdot (\mathbf{H} \times \mathbf{E})dV = -\frac{c}{4\pi} \oint (\mathbf{E} \times \mathbf{H})_n dS_f = 0, \quad (2.4)$$

where $(c/4\pi)(\mathbf{E} \times \mathbf{H})_n$ is a component of the Poynting vector normal to an element of a surface dS_f bounding the field. Here, we used a vector identity $\nabla \cdot (\mathbf{A} \times \mathbf{B}) = \mathbf{B} \cdot \nabla \times \mathbf{A} - \mathbf{A} \cdot \nabla \times \mathbf{B}$ and the Gauss theorem.

After multiplying Eq. (2.4) by time interval δt, it becomes

$$\delta t \int \mathbf{j}_c \mathbf{E}\, dV + \frac{1}{4\pi} \int \mathbf{H}\delta \mathbf{B}\, dV = 0. \qquad (2.5)$$

$$\delta t \int \mathbf{j}_c \mathbf{E}\, dV + \int \mathbf{H}\delta \mathbf{B}\, dV = 0. \qquad \text{(SI)}$$

The first term in (2.5) represents a work done over time δt on the current in the magnet coil $\mathbf{j}_c$ by the electric field $\mathbf{E}$ induced due to changing $\mathbf{B}$ in the system. Therefore, this term taken with opposite sign is the work δW done by the power supply to maintain the set current in the magnet when the induction changes for $\delta \mathbf{B}$, or this is the work needed to change the induction by $\delta \mathbf{B}$, as stated by formula (2.1). Respectively, by virtue of the law of energy conservation, this work is equal to variation of the system energy, referred to as magnetic energy *of the system* [15].

A function of state whose increment in a reversible isothermal process equals the work done on the system is referred to as Helmholtz free energy $F(T, V, \mathbf{B})$.

---

[1] More specifically, it diminishes with rate faster than $1/R^2$ at $R \to \infty$, where R is the distance from the field sources.

If the field with induction **B** is established in the free space, the work done equals energy stored in the field implying that the field internal energy U is

$$U = \int \frac{B^2}{8\pi} dV = \int \frac{H^2}{8\pi} dV$$

$$U = \int \frac{B^2}{2\mu_0} dV = \int \frac{\mu_0 H^2}{2} dV. \quad \text{(SI)}$$

On the other hand, entropy S of the free space is zero by definition (disorder in vacuum is impossible), and therefore the field internal energy equals to its free energy $F = U + TS$. Therefore, $B^2/8\pi = H^2/8\pi$ ($B^2/2\mu_0 = \mu_0 H^2/2$ in SI units) is the density of the Helmholtz free energy of the field in vacuum.

Consider a specimen in free space. In zero field, the Helmholtz potential F_0 of this system (the specimen plus the space around it) equals that of the specimen. For solids[2] infinitesimal increment of F_0 is

$$\delta F_0(T, V) = -S\delta T - P\delta V = -S\delta T,$$

where S is the specimen entropy and P is external pressure assumed zero.

Correspondingly, in the field, the Helmholtz potential of the *system* becomes a function of T, V and **B** so that its increment is

$$\delta F(T, V, \mathbf{B}) = \delta F_0 + \delta W = -S\delta T - P\delta V + \frac{1}{4\pi} \int \mathbf{H}\delta \mathbf{B} \, dV$$

$$(2.6)$$

As known (see Problem 1.1), the magnetic field exerts a force on the specimen surface, so the pressure term in (2.6) now is not zero. In general, the force per unit area due to magnetic field depends on the field magnitude and

---

[2]For a solid body one can assume that pressure of saturated vapor of the body material is zero. Hence, density of the vapor is zero as well. This implies that number of particles N in the body is fixed and therefore the term $\mu_{ch}dN$ in the increment of free energy associated with chemical potential of the body μ_{ch} [26] can be omitted.

its orientation relative to the surface, on possible temperature dependence of μ and other factors. This means that the magnetic field changes not only the volume but also the shape of the body and therefore its demagnetizing factors. All of these make calculations of this term quite complicated even for the cases of the cylindrical and transverse geometries [75]. Fortunately, its contribution to δF of solid specimens of non-magnetostrictive materials is small, so hereinafter this term will be omitted. Note that in this approximation the specimen is incompressible and therefore the Helmholtz free energy $F(T, V, \mathbf{B})$ is identical to the Gibbs free energy $G(T, P, \mathbf{B})$. On that ground, below we will refer to F as merely free energy or thermodynamic potential.

It is more important that the system now consists of the magnetized specimen and the field around it. Therefore, integration in (2.6) has to be carried out over the entire volume occupied by the field. Naturally, this significantly complicates calculation of F, if only because it does not allow one to use specific free energy (free energy density) of the *specimen*, implying that F of such a system is non-additive quantity.

On the first glance, expression for free energy of the specimen can be obtained by subtracting free energy of the applied field from the free energy of the system. This is true, but the outside field caused by the magnetized specimen remains (Problem 1.3) and therefore the specimen energy[3] includes energy of this field, implying that the specimen state functions are still non-additive.

However, the outside field of the specimen depends on its geometry and, as we will see below, for two geometries the account for its contribution is sufficiently simple. Those are the cylindrical and transverse geometries, depicted in Figs. 1.1a and 1.1b. Specimens of these geometries have a primary practical importance, in particular such specimens are used in the vast majority of experiments. Therefore, first we will stop at these two cases.

---

[3] The specimens we consider are motionless, therefore their energy is internal energy U.

2.2 THERMODYNAMICS OF SPECIMENS OF CYLIN-DRICAL GEOMETRY

A longitudinally magnetized uniform cylindrical specimen (a long right cylinder with the base of arbitrary shape in parallel field) can be presented as a sum of thin circular cylindrical elements parallel to the generating line of our specimen. The outer field of each of these elementary cylinders is identical to the outer field of an infinite solenoid (Problem 1.4), i.e., it is zero. Hence, the outer field due to our cylindrical specimen is zero and therefore the field outside it equals to the applied field H_0 all the way down to the specimen surface. Thus, subtracting the free energy of applied field outside the specimen from the free energy of the system, we arrive at the free energy of the specimen which does not contain the outer field. Then the increment of free energy of the *specimen* is

$$\delta F(T, \mathbf{B}) = -S\delta T + \frac{1}{4\pi} \int_V \mathbf{H}\delta\mathbf{B}\, dV, \qquad (2.7)$$

$$\delta F(T, \mathbf{B}) = -S\delta T + \int_V \mathbf{H}\delta\mathbf{B}\, dV, \qquad \text{(SI)}$$

where integration now is carried out over the volume of the specimen V alone.

It is important to note that integral in (2.7) contains variation of (i) the specimen magnetic energy (below we will see that this is mostly change of kinetic energy of electrons) and (ii) energy of the average microscopic field (i.e., of the induction $\mathbf{B}$) in the specimen.

Slightly different justification of formula (2.7) can be as follows [75]. Consider a specimen in the form of a long circular cylinder. A magnet in this case can be a solenoid wound immediately on the specimen. Then, the field inside this magnet is uniform and parallel to the longitudinal axis of the specimen as required for cylindrical geometry. At the same time, the outside field of the specimen is the field outside the solenoid, i.e., zero. So the whole work done by the magnet power supply to change the induction at constant temperature (Eq. (2.1)) goes to the change of the specimen free energy as it takes place in Eq. (2.7).

Based on Eq. (2.7) one can introduce the free energy density of this specimen f (free energy per the specimen unit volume), whose differential is

$$df(T, \mathbf{B}) = -sdT + \frac{\mathbf{H}d\mathbf{B}}{4\pi}, \qquad (2.8)$$

$$df(T, \mathbf{B}) = -sdT + \mathbf{H}d\mathbf{B}, \qquad \text{(SI)}$$

where s is the entropy per unit volume.

However, the free energy $F(T, \mathbf{B})$ or its density $f(T, \mathbf{B})$ is *not* an appropriate state function for the cylindrical geometry because in this case we cannot control $\mathbf{B}$ and therefore cannot use it as independent state parameter. Note that this is not just a matter of inconvenience, as said in some textbooks. Indeed, a quantity which can be calculated from $f(T, \mathbf{B})$ is $\mathbf{H} = 4\pi(\nabla_\mathbf{B} f)_T$[4]. However, the result of this calculation is trivial because in cylindrical geometry $\mathbf{H} = \mathbf{H}_0$ and therefore is known upfront.

Similar situations are well known in traditional thermodynamics. For instance, the Helmholtz free energy density $f(T, V)$ is inappropriate for calculations of thermodynamic properties when controllable parameters of state are temperature and pressure. Appropriate thermodynamic potential in such case is the Legendre transform of the Helmholtz potential referred to as the Gibbs free energy $G(T, P)$ whose density is $g_t = f + PV$ (we designate it g_t to exclude confusion with the linear density of the surface current $\mathbf{g}$).

Apart from $\mathbf{H}_0$ controllable parameters in problems associated with the static magnetic field are current in the magnet

---

[4]Operator $\nabla_\mathbf{A}$ with respect to a vector $\mathbf{A} = A_x\mathbf{i} + A_y\mathbf{j} + A_z\mathbf{k}$ ($\mathbf{i}$, $\mathbf{j}$ and $\mathbf{k}$ are unit vectors along x, y, and z axes, respectively), is defined as

$$\nabla_\mathbf{A} \equiv \frac{\partial}{\partial A_x}\mathbf{i} + \frac{\partial}{\partial A_y}\mathbf{j} + \frac{\partial}{\partial A_z}\mathbf{k}.$$

$\mathbf{j}_c$ and the vector potential due to this current

$$\mathbf{A} = \frac{1}{c}\int \frac{\mathbf{j}_c}{R}dV$$

$$\mathbf{A} = \frac{\mu_0}{4\pi}\int \frac{\mathbf{j}_c}{R}dV, \qquad \text{(SI)}$$

where R is distance from a current element $\mathbf{j}_c dV$ to a point of observation.

As the state parameter, $\mathbf{H}_0$ is more convenient than $\mathbf{j}_c$ or $\mathbf{A}$ because for the use of the latter two quantities one needs to know the current configuration in specific magnet.

Therefore, appropriate and convenient thermodynamic potential(s) in magnetostatics should be function(s) of $\mathbf{H}_0$. Since in the considered geometry all three fields ($\mathbf{H}_0$, $\mathbf{H}$ and $\mathbf{B}$) are parallel, such a potential can be easily constructed via the standard Legendre transform procedure as follows.

Define the free energy $\widehat{F}(T, \mathbf{H})$ so that its density is

$$\widehat{f} = f - \frac{\mathbf{HB}}{4\pi} \qquad (2.9)$$

$$\widehat{f} = f - \mathbf{HB}. \qquad \text{(SI)}$$

Then, using (2.8) we obtain

$$d\widehat{f}(T, \mathbf{H}) = -sdT - \frac{1}{4\pi}\mathbf{B}d\mathbf{H} \qquad (2.10)$$

$$d\widehat{f}(T, \mathbf{H}) = -sdT - \mathbf{B}d\mathbf{H}. \qquad \text{(SI)}$$

Hence, $\widehat{f}$ is function of T and $\mathbf{H}(=\mathbf{H}_0)$ and can be used to calculate specific entropy s

$$s = -\left(\frac{\partial \widehat{f}}{\partial T}\right)_H \qquad (2.11)$$

and $\mathbf{B}$ in specimens of cylindrical geometry

$$\mathbf{B} = -4\pi(\nabla_{\mathbf{H}}\widehat{f})_T = -4\pi(\nabla_{\mathbf{H}_0}\widehat{f})_T \qquad (2.12)$$

$$\mathbf{B} = -(\nabla_{\mathbf{H}}\widehat{f})_T = -\mu_0(\nabla_{\mathbf{B}_0}\widehat{f})_T. \quad \text{(SI)}$$

Due to the parallelism of $\mathbf{H}_0$, $\mathbf{H}$ and $\mathbf{B}$, formula (2.12) takes a simple form

$$B = -4\pi \left(\frac{\partial \widehat{f}}{\partial H}\right)_T = -4\pi \left(\frac{\partial \widehat{f}}{\partial H_0}\right)_T$$

$$B = -\left(\frac{\partial \widehat{f}}{\partial H}\right)_T = -\mu_0 \left(\frac{\partial \widehat{f}}{\partial B_0}\right)_T. \quad \text{(SI)}$$

After calculation of $\mathbf{B}$, the rest of the magnetic properties ($\mathbf{I}$ and $\mathbf{M}$) are computed from definitions (1.9) and (1.2), respectively.

In view of similarity of transformations from $f(T, \mathbf{B})$ to $\widehat{f}(T, \mathbf{H})$ on one side, and from the Helmholtz to the Gibbs potential in traditional thermodynamics on the other, the potential $\widehat{f}(T, \mathbf{H})$ is often referred to as Gibbs potential. This terminology can be confusing because the former procedure is the Legendre transform between two forms of densities of the Helmholtz potential $f(T, V, \mathbf{B})$ and $f(T, V, \mathbf{H}) \equiv \widehat{f}$ in approximation of the constant volume. So, we avoid naming $\widehat{f}$ as the Gibbs potential.

However, $\widehat{F}$ is not the only form of free energy appropriate for the cylindrical geometry. Another thermodynamic potential depending on H_0 can be constructed as follows.

Let us return to the system of the cylindrical specimen plus field, and consider the state function

$$\widetilde{F} = \int \left(\widehat{f} + \frac{H_0^2}{8\pi}\right) dV \qquad (2.13)$$

$$\widetilde{F} = \int \left(\widehat{f} + \frac{B_0^2}{2\mu_0}\right) dV \qquad \text{(SI)}$$

where integration is carried out over volume of the *system*. Note physical meaning of Eq. (2.13): it is free energy $\widehat{F}$ of the system with the specimen minus free energy $\widehat{F}'$ of the system without the specimen $(\widehat{F}' = \int (f' - BH/4\pi)dV = \int (-H_0^2/8\pi)dV)$.

Using (2.10), infinitesimal variation of $\widetilde{F}$ at constant temperature is

$$\delta \widetilde{F} = -\frac{1}{4\pi} \int (\mathbf{B}\delta\mathbf{H} - \mathbf{H}_0\delta\mathbf{H}_0)dV. \qquad (2.14)$$

In the cylindrical geometry $\mathbf{H} = \mathbf{H}_0$, so $\delta\widetilde{F}$ takes form

$$\delta\widetilde{F} = -\frac{1}{4\pi} \int (\mathbf{B} - \mathbf{H})\delta\mathbf{H}_0 dV = -\int \mathbf{I}\delta\mathbf{H}_0 dV, \qquad (2.15)$$

$$\delta\widetilde{F} = -\int (\mathbf{B} - \mu_0\mathbf{H})\delta\mathbf{H}_0 dV = -\int \mathbf{I}\delta\mathbf{B}_0 dV, \qquad (SI)$$

where $\mathbf{I}$ is magnetization defined in Chapter 1.

Remarkably, that integral now is taken over the volume of the specimen alone since outside it $\mathbf{I} = 0$. After Landau and Lifshitz [7], $\widetilde{F}$ is referred to as total free energy of the *specimen*.

In full form, the increment of the total free energy is

$$d\widetilde{F}(T, \mathbf{H}_0) = -SdT - \mathbf{M}d\mathbf{H}_0 \qquad (2.16)$$

$$d\widetilde{F}(T, \mathbf{H}_0) = -SdT - \mu_0\mathbf{M}d\mathbf{H}_0 = -SdT - \mathbf{M}d\mathbf{B}_0, \qquad (SI)$$

where $\mathbf{M}$ is magnetic moment of the specimen induced due to applied field $\mathbf{H}_0$.

The fact that integral in (2.15) is not zero only inside the specimen regardless of the specimen geometry allows one to define the total free energy $\widetilde{F}(T, \mathbf{H}_0)$ for specimens of *any geometry* as

$$\mathbf{M} = -(\nabla_{\mathbf{H}_0}\widetilde{F})_T \qquad (2.17)$$

$$\mathbf{M} = -(\nabla_{\mathbf{B}_0}\widetilde{F})_T, \qquad (SI)$$

Thus, the total free energy of the magnetizing specimen is such a function of state, which derivative with respect to direction of the applied field $\mathbf{H}_0$ equals the specimen magnetic moment taken with opposite sign. Note that this rather cumbersome formulation has a straightforward and simple physical meaning. If $\mathbf{H}_0$ is parallel to one of the axes of the ellipsoidal specimen, then $\mathbf{M}$ is aligned with $\mathbf{H}_0$ (parallel in paramagnetics and antiparallel in diamagnetics). Then Eq. (2.17) takes the form

$$M = -\left(\frac{\partial \widetilde{F}}{\partial H_0}\right)_T. \qquad (2.17a)$$

In general case $\mathbf{H}_0$ should be broken for components parallel to the specimen axes and each component of $\mathbf{M}$ is calculated independently.

As we have seen, for the cylindrical geometry Eq. (2.13) is the correct expression of $\widetilde{F}$, where integral can be immediately taken over the specimen volume because nothing will be changed in transformations from Eq. (2.13) to Eq. (2.15). Therefore, for this geometry we can also introduce a total free energy density

$$d\widetilde{f}(T, \mathbf{H}_0) = -sdT - \mathbf{I}d\mathbf{H}_0, \qquad (2.18)$$

$$d\widetilde{f}(T, \mathbf{B}_0) = -sdT - \mathbf{I}d\mathbf{B}_0, \qquad (\text{SI})$$

and its Legendre transform

$$d(\widetilde{f} + \mathbf{I}\mathbf{H}_0) \equiv df_I(T, \mathbf{I}) = -sdT + \mathbf{H}_0 d\mathbf{I}. \qquad (2.19)$$

$$d(\widetilde{f} + \mathbf{I}\mathbf{B}_0) \equiv df_I(T, \mathbf{I}) = -sdT + \mathbf{B}_0 d\mathbf{I}. \qquad (\text{SI})$$

The last term in this formula can be viewed as work needed to magnetize the specimen unit volume in zero applied field [76]; however, practical importance of $f_I(T, \mathbf{I})$ is the same as that of $f(T, \mathbf{B})$ for cylindrical geometry because $\mathbf{I}$ is uncontrollable quantity.

Using the standard procedure of the Legendre transformation, obtained differential expressions for different forms of free energy density of the specimens of cylindrical geometry can be easily transformed to corresponding expressions for internal energy density $\widehat{u}(s, \mathbf{H})$ and $\widetilde{u}(s, \mathbf{H}_0)$. Specifically,

$$d\widehat{u}(s, \mathbf{H}) \equiv d[\widehat{f}(T, \mathbf{H}) + sT] = Tds - \frac{\mathbf{B}d\mathbf{H}}{4\pi} \qquad (2.20)$$

$$d\widehat{u}(s, \mathbf{H}) \equiv d[\widehat{f}(T, \mathbf{H}) + sT] = Tds - \mathbf{B}d\mathbf{H} \qquad \text{(SI)}$$

and

$$d\widetilde{u}(s, \mathbf{H}_0) \equiv d[\widetilde{f}(T, \mathbf{H}_0) + sT] = Tds - \mathbf{I}d\mathbf{H}_0 \qquad (2.21)$$

$$d\widetilde{u}(s, \mathbf{B}_0) \equiv d[\widetilde{f}(T, \mathbf{B}_0) + sT] = Tds - \mathbf{I}d\mathbf{B}_0. \qquad \text{(SI)}$$

So far in this chapter, we did not assume any specific relation between $\mathbf{B}$ and $\mathbf{H}$. If this relation is linear (i.e., $\mathbf{B} = \mu\mathbf{H}$), then for cylindrical specimens we write

$$\widehat{f}(T, \mathbf{H}) = (\widehat{f})_0 - \frac{\mu H^2}{8\pi} = f_0 - \frac{\mu H_0^2}{8\pi} \qquad (2.22)$$

$$\widehat{f}(T, \mathbf{H}) = (\widehat{f})_0 - \frac{\mu\mu_0 H^2}{2} = f_0 - \frac{\mu B_0^2}{2\mu_0} \qquad \text{(SI)}$$

and

$$\widetilde{f}(T, \mathbf{H}_0) = (\widetilde{f})_0 - \frac{\chi H_0^2}{2} = f_0 - \frac{\chi H_0^2}{2} =$$

$$f_0 + (1 - \mu)\frac{H_0^2}{8\pi}. \qquad (2.23)$$

$$\widetilde{f}(T, \mathbf{H}_0) = f_0 - \frac{\chi^* B_0^2}{2\mu_0} = f_0 + (1 - \mu)\frac{B_0^2}{2\mu_0}. \qquad \text{(SI)}$$

where χ (χ^* in SI units) is magnetic susceptibility defined by (1.15) and $f_0 = F_0/V$ is free energy density at zero field, which is the same for all forms of the magnetic free energy of the specimens with fixed volume.

Finally, the total free energy of specimens of any geometry made of the linear magnetizing materials is

$$\widetilde{F}(T, \mathbf{H}_0) = (\widetilde{F})_0 - \int_0^{H_0} \mathbf{M}d\mathbf{H}_0 = F_0 - \frac{\mathbf{M}\mathbf{H}_0}{2} \qquad (2.24)$$

$$\widetilde{F}(T, \mathbf{B}_0) = (\widetilde{F})_0 - \int_0^{B_0} \mathbf{M}d\mathbf{B}_0 = F_0 - \frac{\mathbf{M}\mathbf{B}_0}{2}. \qquad \text{(SI)}$$

Note that since F_0 and $\mathbf{M} = I V$ are proportional to the specimen volume V, $\widetilde{F}$ in (2.24) is proportional to V too. However, for specimens of other then the cylindrical and (as we will see in the next section) of the transverse geometries $\widetilde{F}/V$ *is not* density of the total free energy because such a "density" includes energy of the outer field caused by the magnetized specimen.

The magnetic term in (2.24), namely

$$E_m = -\int_0^{H_0} \mathbf{M} d\mathbf{H}_0 \qquad (2.25)$$

$$E_m = -\int_0^{B_0} \mathbf{M} d\mathbf{B}_0 \qquad \text{(SI)}$$

is referred to as magnetic energy of the *specimen*. It can be viewed as energy of interaction of the applied magnetic field with the specimen magnetic moment induced by this field.

In electrostatics, an expression similar to (2.25) is work done by the electric field to polarize the specimen which equals potential energy (such as energy of elastic deformation of molecules due to induced electric dipole moment) stored in this specimen plus energy of the electric field caused by the polarized specimen [3]. In magnetostatics E_m is work done by the magnet power supply to magnetize the specimen equal to variation of kinetic energy of the bound electrons due to precession of their orbits[5] induced by the applied magnetic field. In case of non-cylindrical and non-transverse geometries, E_m also includes energy of the outer field created by the magnetized specimen.

Let us look at how it happens in a diamagnetic specimen of cylindrical geometry. For simplicity, we take the specimen consisting of non-interacting single-atomic (hence spherically symmetrical) molecules.

---

[5]In principle, one should also take into account precession of the electron spin; however, its contribution into the variation of kinetic energy is much smaller than that caused by precession of the electron orbits [3].

Under the influence of the magnetic field, electrons in atoms precess with Larmor angular velocity **o** which is

$$\mathbf{o} = -\gamma\mathbf{H},$$

where $\gamma = e/2mc$ ($e/2m$ in SI units) is gyromagnetic ratio of the orbiting electrons.

Hence, linear velocity of each electron $\mathbf{v}_i$ (velocity when no field is applied) changes for $\Delta\mathbf{v}_i = \mathbf{o} \times \mathbf{R}_i$, where $\mathbf{R}_i$ is radius of the electron orbit.

As a result of precession, in accordance with the Lenz law, the sample acquires a negative magnetic moment, the density of which is $\mathbf{I} = \chi\mathbf{H}$, as defined by (1.19). Classical Langevin expression for the magnetic susceptibility χ (see, e.g., [16] for derivation) is

$$\chi = -\frac{NZe^2}{6mc^2}\overline{R^2}, \qquad (2.26)$$

where $\overline{R^2}$ is mean square radius of the electron orbit[6], N is number of molecules per unit volume and Z is the number of electrons in each molecule.

On the other hand, the change of electrons' velocities implies that kinetic energy of electrons in each molecule T_m changes for ΔT_m, which is

$$\Delta T_m = \frac{m}{2}\sum_{i=1}^{Z}[(\mathbf{v}_i + \Delta\mathbf{v}_i)^2 - \mathbf{v}_i^2] = \frac{m}{2}\sum_{i=1}^{Z}[2\mathbf{v}_i\Delta\mathbf{v}_i + (\Delta\mathbf{v}_i)^2].$$

The first term in this equation is

$$m\sum_{i=1}^{Z}\mathbf{v}_i(\mathbf{o} \times \mathbf{R}_i) = m\mathbf{o}\sum_{i=1}^{Z}(\mathbf{R}_i \times \mathbf{v}_i) =$$

$$\mathbf{o}\mathbf{L}_m = -\gamma\mathbf{L}_m\mathbf{H} = -\mathbf{M}_m\mathbf{H},$$

---

[6]The expression for χ in quantum mechanics is the same as Eq. (2.26); the Planck constant enters this formula through calculation of $\overline{R^2}$.

where $\mathbf{L}_m$ and $\mathbf{M}_m$ are the molecular angular and magnetic momenta, respectively.

Hence, since in diamagnetics $\mathbf{M}_m = 0$, we write

$$\Delta T_m = \frac{m}{2} \sum_{i=1}^{Z} (\Delta \mathbf{v}_i)^2 = \frac{m}{2} \sum_{i=1}^{Z} (\mathbf{o} \times \mathbf{R}_i)^2 = \frac{m\gamma^2}{2} \sum_{i=1}^{Z} (\mathbf{H} \times \mathbf{R}_i)^2.$$

In spherically symmetrical atoms the cross-product is (see, e.g., [3, 16] for derivation)

$$(\mathbf{H} \times \mathbf{R}_i)^2 = \frac{2}{3} H^2 R_i^2.$$

Therefore, average change of kinetic energy of electrons per the specimen unit volume (where number of electrons equals NZ) is

$$\Delta e_{kin} = \frac{NZe^2 \overline{R^2}}{12mc^2} H^2,$$

where $\overline{R^2}$ is the same mean square radius of electron orbits as in (2.26). Using χ from (2.26) and taking into account that in the cylindrical geometry $\mathbf{H} = \mathbf{H}_0$, we obtain

$$\Delta e_{kin} = -\frac{1}{2}\chi H^2 = -\frac{1}{2}\mathbf{I}\mathbf{H}_0 = -\frac{1}{V}\int_0^{H_0} \mathbf{M} d\mathbf{H}_0 = \frac{E_m}{V}$$

$$\Delta e_{kin} = -\frac{1}{2}\chi^* \mu_0 H^2 = -\frac{1}{2}\mathbf{I}\mathbf{B}_0 =$$

$$-\frac{1}{V}\int_0^{B_0} \mathbf{M} d\mathbf{B}_0 = \frac{E_m}{V}. \quad \text{(SI)}$$

We see that the specimen magnetic energy and therefore the work done to magnetize the diamagnetic specimen of the cylindrical geometry equals the change of kinetic energy of its electrons. This is also true for paramagnetics with a difference that in paramagnetics $\Delta e_{kin} < 0$ due to non-zero molecular magnetic moments $\mathbf{M}_m$ (see [3] for details).

For the sake of clarity, we used the classical Langevin's approach and a simplified model of the diamagnetic. Magnetism, however, is an essentially quantum phenomenon and therefore more detailed consideration requires a quantum-mechanical approach as demonstrated, e.g., in [21].

Completing this section, we note that the discussed material is close to that of how magnetism in matter is usually presented in textbooks. However, one should remember that all that (except Eq. (2.17)) is related to the specimens of the cylindrical geometry. We underscore that in general the thermodynamic potentials of the magnetizing specimens are non-additive and therefore concept of the density of these potentials as a rule does not make sense. However, there are two important exclusions, the first one is related to the cylindrical geometry we just discussed, and the second exclusion is related to the transverse geometry, which we will discuss next.

2.3 THERMODYNAMICS OF SPECIMENS OF TRANS-VERSE GEOMETRY

A specimen of the transverse geometry is schematically shown in Fig. 1.1b. This is an infinite plane-parallel plate or slab (those with lateral size greatly exceeding its thickness) subjected to a uniform field $\mathbf{H}_0$ applied perpendicular to the the specimen surface.

Consider the specimen in free space. The applied uniform field $\mathbf{H}_0 = \mathbf{B}_0$ (or $\mathbf{H}_0 = \mathbf{B}_0/\mu_0$ in SI units) is the field away from the specimen. Then the symmetry and the flux conservation dictate that the field outside the specimen stays undisturbed all the way up to the specimen surface[7]. Indeed, it cannot bend, because any direction of an imaginary bend is equivalent. Its magnitude cannot change either, because due

---

[7]This does not apply to a spatial range near the surface (so-called "healing length"), where the effects of irregularities of the persistent currents circulating near the surface are significant. In non-superconducting materials, this range is on the order of molecular size. In superconductors, as we will see in the next chapter, it can be significantly greater.

TABLE 2.1 Forms of free energy appropriate for ellipsoidal specimens in *cgs* units. The bottom line in each section is related to the linear magnetizing materials.

Potential	$\eta = 0\ (\mathbf{H} = \mathbf{H}_0)$	$\eta = 1\ (\mathbf{B} = \mathbf{H}_0)$	$0 < \eta < 0$
$F(T, \mathbf{B})$	–	$df = -sdT + \dfrac{\mathbf{H}d\mathbf{B}}{4\pi}$	–
	–	$f = f_0 + \dfrac{B^2}{\mu 8\pi}$	–
$\widehat{F}(T, \mathbf{H})$	$d\widehat{f} = -sdT - \dfrac{\mathbf{B}d\mathbf{H}}{4\pi}$	–	–
	$\widehat{f} = f - \dfrac{BH}{4\pi}$	–	–
	$\widehat{f} = f_0 - \dfrac{\mu H^2}{8\pi}$	–	–
$\widetilde{F}(T, \mathbf{H}_0)$	$d\widetilde{f} = -sdT - \mathbf{I}d\mathbf{H}_0$	$d\widetilde{f} = -sdT - \mathbf{I}d\mathbf{H}_0$	$d\widetilde{F} = -SdT - \mathbf{M}d\mathbf{H}_0$
	$\widetilde{f} = \widehat{f} + \dfrac{H_0^2}{8\pi}$	$\widetilde{f} = f - \dfrac{(1-\mu)}{\mu}\dfrac{H_0^2}{8\pi}$	–
	$\widetilde{f} = f_0 + (1-\mu)\dfrac{H_0^2}{8\pi}$	$\widetilde{f} = f_0 + \dfrac{(1-\mu)}{\mu}\dfrac{H_0^2}{8\pi}$	$\widetilde{F} = F_0 + \dfrac{(1-\mu)}{1-\eta(1-\mu)}\dfrac{VH_0^2}{8\pi}$

TABLE 2.2 Forms of free energy appropriate for ellipsoidal specimens in SI units.

Potential	$\eta = 0$ $(\mathbf{H} = \mathbf{B}_0/\mu_0)$	$\eta = 1$ $(\mathbf{B} = \mathbf{B}_0)$	$0 < \eta < 0$
$F(T, \mathbf{B})$	–	$df = -sdT + \mathbf{H}d\mathbf{B}$ $f = f_0 + \dfrac{B^2}{2\mu\mu_0}$	– –
$\widehat{F}(T, \mathbf{H})$	$d\widehat{f} = -sdT - \mathbf{B}d\mathbf{H}$ $\widehat{f} = f - BH$ $\widehat{f} = f_0 - \dfrac{\mu\mu_0 H^2}{2}$	– – –	– – –
$\widetilde{F}(T, \mathbf{B}_0)$	$d\widetilde{f} = -sdT - \mathbf{I}d\mathbf{B}_0$ $\widetilde{f} = \widehat{f} + \dfrac{B_0^2}{2\mu_0}$ $\widetilde{f} = f_0 + (1-\mu)\dfrac{B_0^2}{2\mu_0}$	$d\widetilde{f} = -sdT - \mathbf{I}d\mathbf{B}_0$ $\widetilde{f} = f - \dfrac{B_0^2}{2\mu_0}$ $\widetilde{f} = f_0 + \dfrac{(1-\mu)}{\mu}\dfrac{B_0^2}{2\mu_0}$	$d\widetilde{F} = -sdT - \mathbf{M}d\mathbf{B}_0$ – $\widetilde{F} = F_0 + \dfrac{(1-\mu)}{1-\eta(1-\mu)}\dfrac{VB_0^2}{2\mu_0}$

to symmetry the field must stay uniform and therefore the change of its magnitude is forbidden by the flux conservation.

Inside the specimen induction $\mathbf{B} = \mathbf{B}_0 = \mathbf{H}_0$ ($\mathbf{B} = \mathbf{B}_0 = \mu_0 \mathbf{H}_0$ in SI units) due to condition (1.16) at the boundary and due to symmetry and the flux conservation at distancing from it.

Therefore, similar to the case of the cylindrical geometry, the thermodynamic potential of the specimen of the transverse geometry is obtained by subtracting the free energy of the applied field outside the specimen from the free energy of the system. This brings us to Eq. (2.7) with a difference that now $\mathbf{B}$ is controllable quantity and therefore $F(T, \mathbf{B})$ is a lawful function of state. Hence one can use the free energy density $f(T, \mathbf{B})$ which increment is expressed by Eq. (2.8). Then the field intensity $\mathbf{H}$ is calculated as

$$\mathbf{H} = 4\pi \nabla_{\mathbf{B}} f = 4\pi \nabla_{\mathbf{H}_0} f \qquad (2.27)$$

$$\mathbf{H} = \mu_0 \nabla_{\mathbf{B}} f = \mu_0 \nabla_{\mathbf{B}_0} f. \qquad \text{(SI)}$$

Having $\mathbf{H}$ and knowing $\mathbf{B} = \mathbf{H}_0$ ($\mathbf{B} = \mathbf{B}_0$ in SI units), one calculates $\mathbf{I} \equiv (\mathbf{B} - \mathbf{H})/4\pi = (\mathbf{H}_0 - \mathbf{H})/4\pi$ and $\mathbf{M} = \mathbf{I}V$.

In diamagnetics $\mathbf{B} < \mathbf{H}$, so in diamagnetic specimens of the transverse geometry $\mathbf{H} > \mathbf{H}_0$. The reversed is true for paramagnetics.

Now we turn to the total free energy. In the transverse geometry $\widehat{F}(T, \mathbf{H})$ is inappropriate for the same reason why $F(T, \mathbf{B})$ is inappropriate for the cylinders. So, we cannot use Eq. (2.13) for the total free energy. An appropriate form of the latter is found as follows.

For the system specimen plus field, consider a function of state

$$\widetilde{F} = \int (f - \frac{B_0^2}{8\pi}) dV = \int (f - \frac{H_0^2}{8\pi}) dV \qquad (2.28)$$

$$\widetilde{F} = \int (f - \frac{B_0^2}{2\mu_0}) dV = \int (f - \frac{\mu_0 H_0^2}{2}) dV, \qquad \text{(SI)}$$

where integral is taken over volume of the system.

Note that $\widetilde{F}$ in (2.28) is free energy F of the system with the specimen minus free energy F' of the system without the specimen $(F' = \int f'dV = \int (H_0^2/8\pi)dV)$.

Using (2.8), variation of $\widetilde{F}$ at constant temperature is

$$\delta\widetilde{F} = \frac{1}{4\pi}\int (\mathbf{H}\delta\mathbf{B} - \mathbf{B}_0\delta\mathbf{B}_0)dV$$

$$\delta\widetilde{F} = \int (\mathbf{H}\delta\mathbf{B} - \frac{\mathbf{B}_0}{\mu_0}\delta\mathbf{B}_0)dV. \quad \text{(SI)}$$

In the transverse geometry $\mathbf{B} = \mathbf{B}_0 = \mathbf{H}_0$ ($\mathbf{B} = \mathbf{B}_0 = \mu_0\mathbf{H}_0$ in SI units), so we rewrite $\delta\widetilde{F}$ as

$$\delta\widetilde{F} = \frac{1}{4\pi}\int (\mathbf{H} - \mathbf{B})\delta\mathbf{B}_0 dV = -\int_V \mathbf{I}\delta\mathbf{H}_0 dV \quad (2.29)$$

$$\delta\widetilde{F} = \int \left(\mathbf{H} - \frac{\mathbf{B}}{\mu_0}\right)\delta\mathbf{B}_0 dV = -\int_V \mathbf{I}\delta\mathbf{B}_0 dV, \quad \text{(SI)}$$

where integral now is taken over volume of the specimen V since outside it $\mathbf{I}$ is zero. Therefore, $\widetilde{F}$ in (2.28) with integral taken over the specimen volume represents the total free energy of the specimen of the transverse geometry.

Upon integration in (2.29), we arrive at the expression identical to Eq. (2.16) which, as we know, is valid for any geometry and automatically fits the definition of total free energy (2.17). Due to that, formulas (2.18) and (2.19) obtained for the cylinders are valid for the transverse geometry as well, but correct form of $\widetilde{F}$ now is given by Eq. (2.28)[8].

We are left to write down the expressions for densities of thermodynamic potentials for the linear magnetizing materials for the case of the transverse geometry. Using (2.8) for f and (2.29) for $\widetilde{f}$ we write

$$f(T,\mathbf{B}) = f_0 + \frac{B^2}{\mu 8\pi} = f_0 + \frac{H_0^2}{\mu 8\pi} \quad (2.30)$$

$$f(T,\mathbf{B}) = f_0 + \frac{B^2}{2\mu\mu_0} = f_0 + \frac{B_0^2}{2\mu\mu_0} \quad \text{(SI)}$$

---

[8] Alternative derivation of Eq. (2.29) is available in [34].

and

$$\tilde{f}(T, \mathbf{H}_0) = f - \frac{H_0^2}{8\pi} = f_0 - \frac{\chi}{\mu} \frac{H_0^2}{2} =$$

$$f_0 + \frac{(1 - \mu)}{\mu} \frac{H_0^2}{8\pi} \quad (2.31)$$

$$\tilde{f}(T, \mathbf{H}_0) = f - \frac{\mu_0 H_0^2}{2} = f_0 - \frac{\chi}{\mu} \frac{\mu_0 H_0^2}{2} =$$

$$f_0 + \frac{(1 - \mu)}{\mu} \frac{B_0^2}{2\mu_0}. \quad (\text{SI})$$

Note the difference in the same thermodynamic potential $\tilde{f}(T, H_0)$ for the cylindrical (Eq. (2.23)) and for the transverse (Eq. (2.31)) geometries. This difference reflects the difference between magnetic moments in these two geometries, as seen from (1.32) and (1.33).

Concluding this section, it should be emphasized that (a) thermodynamic potentials appropriate for specimens of the transverse geometry are $F(T, \mathbf{B})$ and $\tilde{F}(T, \mathbf{H}_0)$, (b) similar to the case of the cylindrical geometry, in transverse geometry the thermodynamic potentials are additive functions of state, (c) the expression of the total free energy $\tilde{F}(T, \mathbf{H}_0)$ in the transverse geometry (Eq. (2.28)) is different from that for the cylindrical geometry (Eq. (2.13)), which reflects the difference of the magnetic moments in these geometries.

2.4 OTHER GEOMETRIES

Thermodynamic potentials for specimens of other than cylindrical and transverse geometries or for the specimens with demagnetizing factor $0 < \eta < 1$ are non-additive due to distortion of the outside field caused by the magnetized specimen. This implies that the concept of the free energy density, as well as the density of any other thermodynamic potential, is inapplicable for these geometries. On the other hand, both $\mathbf{B}$ and $\mathbf{H}$ in these geometries are out of our control and therefore neither $F(T, \mathbf{B})$ nor $\widehat{F}(T, \mathbf{H})$ can be used for description of thermodynamic properties. As always, controllable parameter of state is the applied field $\mathbf{H}_0$ and therefore the total free energy $\widetilde{F}(T, \mathbf{H}_0)$ is the only appropriate kind of free energy. We remind that $\widetilde{F}$ is defined by Eq. (2.17) and its differential is

$$d\widetilde{F} = -SdT - \mathbf{M}d\mathbf{H}_0. \tag{2.16}$$

Using definitions of $\mathbf{I}$ (1.2) and $\mathbf{H}$ (1.9), Eq. (2.16) is rewritten as

$$d\widetilde{F} = -SdT - \frac{1}{4\pi} \int_V (\mathbf{B} - \mathbf{H})\delta\mathbf{H}_0 dV \tag{2.32}$$

$$d\widetilde{F} = -SdT - \int_V (\frac{\mathbf{B}}{\mu_0} - \mathbf{H})\delta\mathbf{B}_0 dV. \tag{SI}$$

These two formulas (Eqs. (2.16) and (2.32)) represent the thermodynamic potential appropriate for any specimen at any orientation of the applied field $\mathbf{H}_0$ regardless of specific form of relationship between $\mathbf{B}$ and $\mathbf{H}$.

For the linear magnetizing materials, magnetic moment of any ellipsoidal specimen is (see Problem 1.10)

$$\mathbf{M} = -\frac{V}{4\pi} \frac{(1 - \mu)}{1 - \eta(1 - \mu)} \mathbf{H}_0. \tag{p1.10-6}$$

$$\mathbf{M} = -\frac{(1 - \mu)V}{1 - \eta(1 - \mu)} \frac{\mathbf{B}_0}{\mu_0}. \tag{SI}$$

Therefore, $\widetilde{F}$ for this kind of materials is

$$\widetilde{F} = F_0 + \frac{(1 - \mu)}{1 - \eta(1 - \mu)} \frac{V\mathbf{H}_0^2}{8\pi} \tag{2.33}$$

$$\widetilde{F} = F_0 + \frac{(1 - \mu)}{1 - \eta(1 - \mu)} \frac{V\mathbf{B}_0^2}{2\mu_0}, \tag{SI}$$

where F_0 is the specimen free energy at zero field.

We stress that $(\partial \widetilde{F}/\partial V)_{H_0}$ is density of the total free energy for the specimens of the cylindrical ($\eta = 0$) and the transverse ($\eta = 1$) geometries *only*; for other geometries ($0 < \eta < 1$) $\widetilde{F}/V$ or $(\partial \widetilde{F}/\partial V)_{H_0}$ does not have a physical meaning. For cylindrical geometry ($\eta = 0$) Eq. (2.33) yields

$$\widetilde{F} \equiv (\widetilde{F})_\| = F_0 + (1 - \mu)\frac{V\mathbf{H}_0^2}{8\pi} \qquad (2.34)$$

which is fully consistent with Eq. (2.23).

And $\widetilde{F}$ for the transverse geometry ($\eta = 1$) following from Eq. (2.33) is

$$\widetilde{F} \equiv (\widetilde{F})_\perp = F_0 + \frac{(1 - \mu)}{\mu}\frac{V\mathbf{H}_0^2}{8\pi}, \qquad (2.35)$$

which is the same as that in Eq. (2.31).

This confirms validity of Eq. (2.33) and serves as an additional confirmation of the Poisson theorem and Eq. (1.22) which is based on it.

Forms of free energy appropriate for ellipsoidal specimens of different geometries are summarized in Tables 2.1 and 2.2.

We conclude this chapter by noting that for regular dia- and paramagnetics in which $|1 - \mu| \ll 1$, differences between the different forms of the free energy are small. However, in case of superconducting materials the use of inappropriate function of state leaves no chance for any plausible result.

2.5   PROBLEMS

2.1. Find temperature dependence of entropy of paramagnetic specimens of cylindrical geometry obeying the Curie law $\chi = C/T$, where C is a constant of material (Curie constant).

Solution. In cylindrical geometry, $\mathbf{M}$ is parallel to $\mathbf{H}_0$, so we drop the vector sign and write

$$d\widetilde{F} = -SdT - MdH_0.$$

Hence,

$$S = -\left(\frac{\partial \widetilde{F}}{\partial T}\right)_{H_0}$$

and

$$M = -\left(\frac{\partial \widetilde{F}}{\partial H_0}\right)_T.$$

Therefore

$$\left(\frac{\partial S}{\partial H_0}\right)_T = \left(\frac{\partial M}{\partial T}\right)_{H_0} = V\left(\frac{\partial(\chi H)}{\partial T}\right)_{H_0} = -\frac{CVH_0}{T^2}.$$

So,

$$S(T, H_0) = S(T, H) = S_0(T) - \frac{CVH_0^2}{2T^2} = S_0(T) - \frac{CVH^2}{2T^2}.$$

2.2. Show that magnetic entropy S_m (i.e., magnetic part of the specimen entropy) of diamagnetics is zero. Use H_0 parallel to the specimen axis.

Solution. At the given field orientation, (2.16) is rewritten as

$$d\widetilde{F} = -SdT - MdH_0.$$

In diamagnetics χ and therefore M is temperature independent, so

$$\left(\frac{\partial S}{\partial H_0}\right)_T = \left(\frac{\partial M}{\partial T}\right)_{H_0} = 0.$$

Therefore, the specimen entropy does not depend on the field and equals $S = S_0(T)$, entropy at zero field. In its turn $S_0 = S_{lat} + S_{m0}$, where S_{lat} is the lattice part of the entropy and S_{m0} is its magnetic part associated with disorder in orientation of molecular magnetic moments in zero field. In diamagnetics, net magnetic moment of each molecule in zero field is zero by definition, which automatically implies that $S_{m0} = 0$ as well.

Alternatively, the absolute value of S_m can be obtained from statistical mechanics, where entropy of a system is

$$S = k_B \ln \Omega.$$

Here k_B is the Boltzmann constant and Ω is the number of accessible states of the system. If the system consists of N units, each of which can be with equal probability found in n states, then $\Omega = n^N$.

The magnetic component of the diamagnetic specimen (of the ellipsoidal shape) in the uniform applied field represents a system of identical (both in magnitude and direction) induced magnetic moments of its N molecules. Therefore, the number of accessible states for the magnetic moment of each molecule is 1 and therefore $\Omega = 1^N = 1$.

Hence,

$$S_m = k_B \ln 1 = 0.$$

2.3. Find entropy of a paramagnetic specimen of cylindrical geometry with given $\mu(T)$.

Solution.

Since $\mathbf{M} = I\mathbf{V}$ is parallel to $\mathbf{H}_0$, we may drop vector signs and write $\widetilde{f}$, defined by (2.18), as

$$d\widetilde{f} = -sdT - IdH_0.$$

Then, taking into account that $H = H_0$, we obtain

$$\left(\frac{\partial s}{\partial H} \right)_T = \left(\frac{\partial I}{\partial T} \right)_H = H \left(\frac{\partial \chi}{\partial T} \right)_H = \frac{H}{4\pi} \frac{\partial \mu}{\partial T}. \quad \text{(p. 2.3-1)}$$

We keep notation of the partial derivative $\partial \mu / \partial T$ as a reminder that it should be taken at constant volume. Integrating Eq. (p. 2.3-1) over H at constant T, we get

$$s = s_0(T) + \frac{\partial \mu}{\partial T} \frac{H^2}{8\pi}.$$

Alternatively, one can calculate s as $s = -(\partial \widehat{f}/\partial T)_H$ (see (2.10)) using $\widehat{f}$ from (2.22). This yields the same formula

$$s = -\frac{\partial f_0}{\partial T} + \frac{\partial \mu}{\partial T} \frac{H^2}{8\pi} = s_0(T) + \frac{\partial \mu}{\partial T} \frac{H^2}{8\pi}. \qquad \text{(p. 2.3-2)}$$

And one more way to calculate s is through direct use of (2.18) and (2.23):

$$s = -\left(\frac{\partial \widetilde{f}}{\partial T}\right)_T = -\frac{\partial f_0}{\partial T} + \frac{\partial \mu}{\partial T} \frac{H^2}{8\pi} = s_0(T) + \frac{\partial \mu}{\partial T} \frac{H^2}{8\pi}.$$

So, all three approaches yield the same answer and if $\chi = C/T$, one arrives at the answer found in Problem 2.1. Note that magnetic part of entropy is determined by derivative $\partial \mu/\partial T$. In diamagnetics, this part is zero in consistence with the result of the previous problem.

2.4. Find entropy of a paramagnetic specimen of the transverse geometry. Express entropy as function of H_0 and H.

Solution. Using the free energy density $f(T, \mathbf{B})$ (2.8) and $\widetilde{f}(T.\mathbf{H}_0)$ (2.18), and the fact that $H_0 = B$, we write

$$\left(\frac{\partial s}{\partial H}\right)_T = \left(\frac{\partial I}{\partial T}\right)_{H_0} = \left(\frac{\partial \chi H}{\partial T}\right)_{H_0} = \frac{\partial}{\partial T}\left(\frac{\mu - 1}{4\pi}H\right)_{H_0} =$$

$$-\frac{\partial}{\partial T}\left(\frac{H_0}{4\pi\mu}\right)_{H_0} = \frac{H_0}{4\pi}\left(\frac{1}{\mu^2}\frac{\partial \mu}{\partial T}\right).$$

Therefore,

$$s(T, H_0) = s(T, B) = s_0(T) + \frac{1}{\mu^2}\left(\frac{\partial \mu}{\partial T}\right)\frac{B^2}{8\pi} =$$

$$s_0(T) + \frac{1}{\mu^2}\left(\frac{\partial \mu}{\partial T}\right)\frac{H_0^2}{8\pi} = s_0(T) + \left(\frac{\partial \mu}{\partial T}\right)\frac{H^2}{8\pi}.$$

We see that, in the transverse geometry, $S(H_0)$ differs from that in the cylindrical geometry (Problem 2.3), but $S(H)$ (i.e., degree of disorder vs. internal field H) is the same.

Similarly, as it was done in the previous problem, the same expression for $s(T, H_0)$ can be obtained by direct differentiating of Eq. (2.8) and Eq. (2.18).

2.5. Given is an ellipsoidal specimen of constant volume V made of a paramagnetic material with permeability $\mu(T)$. The specimen is in a magnetic field H_0 oriented parallel to one of the ellipsoidal axes; the demagnetizing factor with respect to this axis is η. (a) Find the specimen entropy as functions of H_0 and H and show that $S(H)$ does not depend on η. (b) Is this formula consistent with results found in Problems 2.3 and 2.4? Justify your answer.

Hint: Use the result of Problem 1.10.

Answer:

$$(a) \quad S(T, H_0) = S_0(T) + \frac{V}{[1 - \eta(1 - \mu)]^2} \left(\frac{\partial \mu}{\partial T}\right) \frac{H_0^2}{8\pi},$$
$$(\text{p. 2.5-1})$$

$$S(T, H) = S_0(T) + V \left(\frac{\partial \mu}{\partial T}\right) \frac{H^2}{8\pi}, \qquad (\text{p. 2.5-2})$$

where S_0 is the specimen entropy in zero field.

Note that in different geometries (different η) $S(H_0)$ is different function of μ. However, S vs. H is the same for all specimen-field configurations. This is one more demonstration of importance of the field H in magnetostatics.

(b) Yes: for η equal zero and unity Eq. (p. 2.5-1) converts to the answers of Problems 2.3, and 2.4, respectively.

2.6. Show that the total free energy $\widetilde{F}$ at constant temperature for any ellipsoidal specimen with a fixed (field independent) magnetic susceptibility χ in the magnetic field H_0 parallel to one of the specimen axes, i.e., $\widetilde{F}$ of a specimen with any demagnetizing factor $0 \leq \eta \leq 1$, can be expressed by a simple formula

$$\widetilde{F} = F_0 - \chi V \frac{H H_0}{2},$$

where F_0 is the specimen free energy at zero field, V is the specimen volume and H is the field intensity inside the specimen.

Solution

$$\widetilde{F} = F_0 - V \int_0^{H_0} I dH_0.$$

Using Eq. (1-22), we obtain

$$H(1 + 4\pi\eta\chi) = H_0.$$

So,

$$\int_0^{H_0} I dH_0 = \chi \int_0^{H_0} H dH_0 = \chi \frac{H_0^2}{2(1 + 4\pi\eta\chi)} = \chi \frac{H H_0}{2}.$$

2.7. In the previous problem, apply the field H_0 in arbitrary direction and show that $\widetilde{F}$ is

$$\widetilde{F} = F_0 - \chi V \frac{\mathbf{H} \cdot \mathbf{H_0}}{2}.$$

2.8. Show that in paramagnetics, obeying the Curie law ($\chi = C/T$), entropy is a function of H/T and that this fact leads to the magnetocaloric effect. Consider specimen of the cylindrical geometry. In which conditions does this effect have a practical significance?

Solution.

For cylindrical geometry formula (p. 2.5-1) is rewritten as

$$S(T, H) = S_0(T) + V \left(\frac{\partial \chi}{\partial T}\right) \frac{H^2}{2} = S_0(T) - V \frac{C}{2} \left(\frac{H_0}{T}\right)^2$$

The first term (S_0) is a lattice part of the specimen entropy. If this term is small compared to the second term, representing the magnetic part of the entropy, the constancy of the specimen entropy S implies the constancy of H_0/T. Or at constant S

$$T_f = \frac{H_{0f}}{H_{0i}} T_i,$$

where T_i and T_f are the specimen temperatures at the initial H_{0i} and the final H_{0f} applied fields, respectively.

Therefore, in accordance with the magnetocaloric effect [24], increase of the field leads to increase of the specimen temperature and vice versa. To have practical significance, the magnetic part of entropy must be dominant. This is possible at temperatures much less than the Debye temperature.

2.9. For a paramagnetic specimen with given $\mu(T)$, calculate (a) C_H, heat capacity at constant H, and (b) C_B, heat capacity at constant B. (c) Find C_H and C_B for paramagnetics obeying the Curie law ($\chi = C/T$).

Solution.

Since the field H is controllable in specimens of the cylindrical geometry, this geometry should be used to calculate C_H. Respectively, the transverse geometry should be used to calculate C_B. So, using solutions of Problems 2.3 and 2.4, one obtains

$$C_H \equiv (C_H)_\parallel = T \left(\frac{\partial S_\parallel}{\partial T} \right)_H =$$
$$C_0 + \frac{TH^2}{8\pi} V \frac{\partial^2 \mu}{\partial T^2} = C_0 + \frac{TH_0^2}{8\pi} V \frac{\partial^2 \mu}{\partial T^2}$$

and

$$C_B \equiv (C_B)_\perp = T \left(\frac{\partial S_\perp}{\partial T} \right)_H =$$
$$C_0 + \frac{TB^2}{8\pi\mu^2} V \left[\frac{\partial^2 \mu}{\partial T^2} - \frac{2}{\mu} \left(\frac{\partial \mu}{\partial T} \right)^2 \right] =$$
$$C_0 + \frac{TH_0^2}{8\pi\mu^2} V \left[\frac{\partial^2 \mu}{\partial T^2} - \frac{2}{\mu} \left(\frac{\partial \mu}{\partial T} \right)^2 \right],$$

where C_0 is the specimen heat capacity at constant volume and zero field, and H_0 is the applied field.

In the case of validity of the Curie law, these formulas are

$$C_H \equiv (C_H)_\parallel = C_0 + CV \left(\frac{H}{T}\right)^2 = C_0 + CV \left(\frac{H_0}{T}\right)^2$$

and

$$C_B \equiv (C_B)_\perp = C_0 + \frac{CV}{\mu^3} \left(\frac{B}{T}\right)^2 = C_0 + \frac{CV}{\mu^3} \left(\frac{H_0}{T}\right)^2.$$

So, in paramagnetics $C_H > C_B$, whereas heat capacity of diamagnetics does not depend on the field, i.e, $C_H = C_B = C_0$.

2.10. Calculate the change of temperature of the paramagnetic specimens of (a) cylindrical and (b) transverse geometries caused by an adiabatic change of the applied field H_0. Take $\chi(T) \equiv (\mu - 1)/4\pi = C/T$.

Solution.
(a) Cylindrical geometry.
For densities of free energy $\widehat{f}$ and entropy s we write

$$d\widehat{f} = -sdT - \frac{BdH}{4\pi}.$$

Hence,

$$\left(\frac{\partial s}{\partial H}\right)_T = \frac{1}{4\pi} \left(\frac{\partial B}{\partial T}\right)_H$$

In the adiabatic process $s = const$, therefore

$$ds(T, H) = \left(\frac{\partial s}{\partial T}\right)_H dT + \left(\frac{\partial s}{\partial H}\right)_T dH =$$

$$\frac{1}{T} \left(\frac{T\partial s}{\partial T}\right)_H dT + \frac{1}{4\pi} \left(\frac{\partial(\mu H)}{\partial T}\right)_H dH = \frac{c_H}{T} dT + \frac{H}{4\pi} \frac{\partial \mu}{\partial T} dH = 0$$

Therefore, turning to finite increments, the temperature change ΔT is

$$\Delta T = -\frac{TH}{4\pi C_H} V \frac{\partial \mu}{\partial T} \Delta H = THV \frac{C}{T^2} \left[C_0 + CV \left(\frac{H}{T}\right)^2\right]^{-1},$$

where C is the Curie constant, and C_H and C_0 are the heat capacities of the specimen at constant field H and at zero field, respectively.

If initial specimen temperature T is much less than the Debye temperature T_D, C_0 can be dropped and the temperature change is

$$\Delta T = (\Delta T)_\| = \frac{T}{H}\Delta H = \frac{T}{H_0}\Delta H_0,$$

where T and $H = H_0$ are initial temperature and the applied field, and ΔT is the temperature change caused by the field change ΔH_0.

Consistency of this formula with the solution of Problem 2.8 is obvious.

(b) Transverse geometry.

For the transverse geometry, controllable field is $B = H_0$ and the appropriate free energy is $F(T, B)$. Entropy $S = S(T, B)$ and the heat capacity $C_B(T, B)$ were calculated in Problems 2.4 and 2.9.

Performing steps similar to those above, one arrives at

$$\Delta T = -\frac{TB}{4\pi C_B \mu^2}V\frac{\partial\mu}{\partial T}\Delta B = \frac{TBV}{\mu^2}\frac{C}{T^2}\left[C_0 + \frac{CV}{\mu^3}\left(\frac{B}{T}\right)^2\right]^{-1}.$$

At $T \ll T_D$, the temperature change in the transverse geometry $(\Delta T)_\perp$ is

$$\Delta T = (\Delta T)_\perp = \frac{T\mu}{B}\Delta B = \frac{T\mu}{H_0}\Delta H_0.$$

Comparing $(\Delta T)_\perp$ with $(\Delta T)_\|$ we see that the magnetocaloric effect is more significant in the transverse geometry.

CHAPTER 3

Thermodynamics of Superconductors

3.1 CONDENSATION ENERGY

The key thermodynamic quantity distinguishing superconducting state from any other known state of matter is condensation energy E_c defined in (1.35). E_c represents energy advantage of the superconducting state with respect to the normal state of a specimen at the same temperature in zero field. The last circumstance immediately leads to the fact that E_c is an additive quantity (proportional to the specimen volume V) independent on the specimen shape. Hence, one can rewrite (1.35) as

$$E_c(T, V) \equiv F_{n0}(T, V) - F_{s0}(T, V) = e_c(T)V, \qquad (3.1)$$

where condensation energy density e_c is function of temperature only.

In turn, e_c defines the thermodynamic critical field H_c via expression (1.36). Therefore, H_c is a function of temperature directly associated with E_c. According to experimental data, in many materials dependence of H_c on T is close to parabolic [11, 35, 54]

$$H_c(T) = H_{c0}(1 - (T/T_c)^2), \qquad (3.2)$$

where T_c is critical temperature at zero field and H_{c0} is critical field at zero temperature. In high-purity specimens T_c and H_{c0} are macroscopic intrinsic parameters of the superconducting material determined by its chemical composition[1] and crystalline structure; as for the rest, T_c and H_{c0} can vary depending on chemical and structural defects in specific material[2].

On the other hand, in zero field free energies $F(T, V, \mathbf{B})$, $\widehat{F}(T, V, \mathbf{H})$ and $\widetilde{F}(T, V, \mathbf{H}_0)$ are identical, so

$$E_c(T, V) \equiv F_{n0} - F_{s0} = \widehat{F}_{n0} - \widehat{F}_{s0} = \widetilde{F}_{n0} - \widetilde{F}_{s0}. \quad (3.3)$$

Therefore, using (1.36), general expression for the total free energy at fixed temperature (2.24) for the case of a superconducting specimen takes the form

$$\widetilde{F} = \widetilde{F}_{s0} - \int_0^{H_0} \mathbf{M} d\mathbf{H}_0 = \widetilde{F}_n - \frac{H_c^2}{8\pi} V - \int_0^{H_0} \mathbf{M} d\mathbf{H}_0 \quad (3.4)$$

$$\widetilde{F} = \widetilde{F}_{s0} - \int_0^{B_0} \mathbf{M} d\mathbf{B}_0 = \widetilde{F}_n - \frac{B_c^2}{2\mu_0} V - \int_0^{B_0} \mathbf{M} d\mathbf{B}_0, \quad (\text{SI})$$

where $\widetilde{F}_{n0}$ is replaced by $\widetilde{F}_n$ since the total free energy in the normal state does not depend on the field (because $\mathbf{M}$ in the N state is zero by definition). Important to underscore that (3.4) is valid for single-connected superconducting specimens of any geometry.

Since in superconductors $\mathbf{M}$ is negative, from (3.4) we see that $\widetilde{F}$ in the S state, being less than $\widetilde{F}_n$, increases with increasing the applied field H_0. This implies that above a

---

[1]In some metals T_c also depends on the isotope composition [29].

[2]The level of material contamination can be characterized by the residual resistivity ratio (RRR), the ratio of resistivity at room temperature to its residual value at $T \to 0$. RRR is determined by the mean free path of electrons. Normally, materials with RRR≈300-500 and higher are considered as pure. However *there is no* fixed RRR number dividing pure and impure materials and the ultimate test of material purity is reproducibility of $M(H_0)$ (or another *bulk* magnetic property, e.g., μSR measured induction) data obtained with the specimen cooled in zero field (abbreviated as ZFC standing for "in zero-field cooled") and in the field well exceeding ($> H_{c3}$) the S/N critical field (abbreviated as FC).

certain critical field H_{cr} the S state inevitably becomes thermodynamically unfavorable and at $H_0 = H_{cr}$ the superconducting specimen undergoes transition to the N state and vice versa. Hence, reversible S/N transition, which is an experimentally established obligatory property of all superconductors, is a direct consequence of existence of the finite condensation energy.

At the critical field $\widetilde{F}(H_{cr}) = \widetilde{F}_n$ and therefore the specimen magnetic energy equals its condensation energy, i.e.

$$- \int_0^{H_{cr}} \mathbf{M} d\mathbf{H}_0 = \frac{H_c^2}{8\pi} V. \qquad (1.37)$$

Expressing $\mathbf{M}$ and $\mathbf{H}_0$ in dimensionless units $\mathbf{M}' = 4\pi\mathbf{M}/VH_c$ and $\mathbf{H}_0' = \mathbf{H}_0/H_c$ (those are $\mathbf{M}' = \mathbf{M}\mu_0/VB_c$ and $\mathbf{B}_0' = \mathbf{B}_0/B_c$ in SI units), Eq. (1.37) is rewritten as

$$- \int_0^{H_{cr}'} \mathbf{M}' d\mathbf{H}_0' = - \int_0^{H_{cr}} \left(\frac{4\pi\mathbf{M}}{VH_c} \right) \frac{d\mathbf{H}_0}{H_c} = \frac{1}{2} \qquad (3.5)$$

$$- \int_0^{B_{cr}'} \mathbf{M}' d\mathbf{B}_0' = - \int_0^{B_{cr}} \left(\frac{\mu_0\mathbf{M}}{VB_c} \right) \frac{d\mathbf{B}_0}{B_c} = \frac{1}{2}. \qquad (\text{SI})$$

Eq. (1.37) allows one to determine E_c from experimental magnetization curves $\mathbf{M}$ vs. $\mathbf{H}_0$ ($\mathbf{M}$ vs. $\mathbf{B}_0$ in SI units) at constant temperatures. If $\mathbf{H}_0$ is parallel to one of the specimen axes, then $\mathbf{M}$ is aligned with $\mathbf{H}_0$ and E_c equals to area under the magnetization curve. Change of direction of $\mathbf{H}_0$ changes the magnetization curve in such a way that the integral in (1.37) stays constant. If the magnetization curve is graphed in coordinates M' vs. H', then the area under this graph is $1/2$ regardless of the specimen material, temperature and the field orientation, provided it is parallel to one of the specimen axes. This law is referred to as a rule of $1/2$. As usual, in case of arbitrary orientation of $\mathbf{H}_0$, it should be broken for components parallel to the specimen axes. Then the specimen magnetic energy in Eq. (3.4) and condensation energy in Eq. (1.37) is the sum of corresponding quantities for each component.

All these facts were well established long ago [35]. Let us now ask: what do they mean? The answer provided by thermodynamics is as follows.

Condensation energy is an internal resource allowing the specimen to stay superconducting in magnetic field. In diamagnetics the change of free energy at constant temperature (which is the specimen magnetic energy) equals the change of internal energy. Comparing (3.4) and (2.24) we see, that in contrast to the normal diamagnetics a source of the magnetic energy of superconductors is condensation energy. In particular, as seen from Eq. (1.37), maximum specimen magnetic energy or maximum change of its internal energy at constant temperature equals the condensation energy available at this temperature at the beginning of the journey along the field (at $H_0 = 0$). Therefore, Eq. (1.37) and following from it the rule of $1/2$ represent the law of energy conservation or the first law of thermodynamics for superconductors. On the other hand, the fact that at thermodynamic equilibrium the free energy of any system takes a minimum of its possible values implies that the condensation energy in superconductors is consumed with the maximum possible effectiveness or with maximum possible saving.

Owing to the condensation energy, superconductor is free to set currents necessary for preservation of the S state in magnetic field regardless of whether there is or there is no flux change in the system, in complete coherence with the Meissner effect. Thus, compliance with the rule of $1/2$ following from the existence of the condensation energy, is the necessary condition of justification of experimental results (i.e., whether the specimen used in an experiment can be regarded as a superconductor or as a perfect conductor) and of validity of a theory addressing the equilibrium properties.

3.2   TYPE-I SUPERCONDUCTORS

Typical phase diagram of type-I superconductors is shown in Fig. 3.1. The curve labeled as "parallel field" represents the temperature dependence of the thermodynamic critical field $H_c(T)$ obtained from measurements of magnetization on high-purity indium specimens of different forms and sizes in the parallel field; demagnetizing factor of all these specimens is close to zero [34, 77]. These specimens are in the Meissner state at (T, H_0) below this curve, and in the N state at (T, H_0) lying above it. The lower curve represents the perpendicular component of the S/N critical field $H_{cr}(T)$ determined from photo-magnetic images and measurements of electrical resistance on 2.5 μm-thick film in a tilted field [34]. Locus of this curve depends on the specimen thickness d. In thick specimens (a few millimeters and thicker[3]), the lower curve merges with the upper one. At the state parameters below the lower curve, the specimen of the transverse geometry is in the intermediate state, and it is in the N state at the parameters above this curve. Note, at (T, H_0) between the curves, the specimen in the transverse field is in the N state in spite of the fact that H_0 is less than H_c.

This and other equilibrium properties of type-I superconductors are considered below.

3.2.1 Cylindrical geometry

We start from the specimens of cylindrical geometry[4]. Such specimens are in the Meissner state over the entire field and temperature range of the S state, hence they always represent perfect diamagnetics ($\mu = 0$). Typical magnetization curves for these specimens are shown in Fig. 1.2. Magnetic moment in this case is $\mathbf{M} = -\mathbf{H}_0 V/4\pi$ (see Eq. (1.32)). Plugging it

---

[3]For 1-mm-thick indium disc the difference between H_{cr} measured in parallel and perpendicular field is $\approx 5\%$ (V.K., not published)

[4]The specimens are assumed as sufficiently massive, implying that their minimal dimension is much greater than the penetration depth λ. For reference, at low temperatures typical value of λ in type-I superconductors is on the order of 10^{-6} cm [78]; in type-II materials it starts from similar value in Nb [79] and increases for an order of magnitude in high-T_c materials [80, 81].

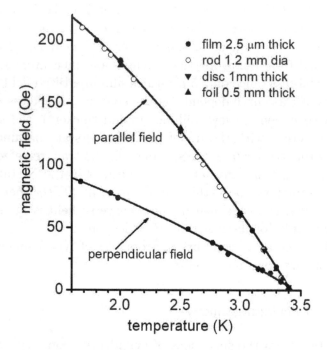

FIGURE 3.1   Phase diagram of pure indium. Experimental data shown by open circles are taken from Ref. [77]; the rest of the data are from Ref. [34]; solid curves are parabolic fits of the experimental data. The upper curve represents thermodynamic critical field $H_c(T)$, the lower curve is $H_{cr}(T)$ for 2.5-μm-thick film in perpendicular field. (Reprinted from [34] with permission from the American Physical Society.)

in Eq. (1.37) one finds that $H_{cr} = H_c$. As already noted, this equality is valid for the cylindrical geometry only.

Eq. (3.4) for this geometry is written as

$$\widetilde{F} = \widetilde{F}_n - \frac{H_{cr}^2}{8\pi}V + \frac{H_0^2}{8\pi}V. \qquad (3.6)$$

A specimen entropy in the superconducting state S_s is

$$S_s = -\left(\frac{\partial \widetilde{F}}{\partial T}\right)_{H_0} = S_n + \frac{H_{cr}V}{4\pi}\frac{dH_{cr}}{dT} \qquad (3.7)$$

$$S_s = -\left(\frac{\partial \widetilde{F}}{\partial T}\right)_{H_0} = S_n + \frac{B_{cr}V}{\mu_0}\frac{dB_{cr}}{dT}, \qquad \text{(SI)}$$

where $S_n = -(\partial \widetilde{F}_n/\partial T)_{H_0}$ is entropy in the N state, which does not depend on the field because $\mu(=1)$ is fixed by definition.

Since H_{cr} decreases with increasing temperature, from (3.6) we see that entropy S_s is less than S_n at the same temperature. This means that motion of electrons in the superconducting state is more ordered than that in the N state (electrons are coupled and move in pairs).

On the other hand, since the specimen magnetic moment in the S state at constant H_0 does not depend on temperature, S_s does not depend on the field[5]. Therefore, $\Delta S = S_n - S_s$ in (6) equals the difference of entropies of the N and S states along the coexisting curve $H_{cr}(T)$ or the jump of the specimen entropy at the S/N transition.

Multiplying ΔS by T, we find the latent heat of the S/N transition Q, which is

$$Q = -TV\frac{H_{cr}}{4\pi}\frac{dH_{cr}}{dT} = -TV\frac{H_c}{4\pi}\frac{dH_c}{dT} \qquad (3.8)$$

---

[5]This follows from the Maxwell relation $\left(\frac{\partial S}{\partial H_0}\right)_T = \left(\frac{\partial M}{\partial T}\right)_{H_0}$ and from the fact that the magnetic permeability $\mu(=0)$ is constant (Problem 2.1).

$$Q = -TV \frac{B_{cr}}{\mu_0} \frac{dB_{cr}}{dT} = -TV \frac{B_c}{\mu_0} \frac{dB_c}{dT}. \qquad \text{(SI)}$$

This formula was for the first time proposed by W. Keesom at the Fourth Solvay conference in discussion following a report of Kamerlingh Onnes on superconductivity [82]. Q is the heat absorbed at the isothermal transition from the S to the N state and released at the transition in opposite direction. The latent heat is the hallmark of the first order phase transition, so its existence, which was confirmed in many experiments (see [35]), serves as an additional confirmation that the S/N transition in type-I superconductors is the first order phase transition. In Chapter 1 we deduced this fact from the positive S/N interphase surface energy. Another hallmark of the first order phase transition is hysteresis associated with possibility of the metastable state. This hysteresis is well seen in Figs. 1.2 and 1.4[6].

Differentiating Eq. (3.7) with respect to T and multiplying by T, we find the difference of heat capacities along the coexistence curve ΔC, which is

$$\Delta C = C_n - C_s = T \frac{dS_n}{dT} - T \frac{dS_s}{dT} =$$
$$- \frac{TV}{4\pi} \left[H_{cr} \frac{d^2 H_{cr}}{dT^2} + \left(\frac{dH_{cr}}{dT} \right)^2 \right] \qquad (3.9)$$

$$\Delta C = C_n - C_s = - \frac{TV}{\mu_0} \left[B_{cr} \frac{d^2 B_{cr}}{dT^2} + \left(\frac{dB_{cr}}{dT} \right)^2 \right]. \qquad \text{(SI)}$$

In the upper end point of the coexistence curve (at $T = T_c$ and $H_0 = 0$) the entropy jump ΔS along with the latent heat

---

[6]In some textbooks it is said that the metastable states exist on the both sides of the S/N transition. However experimenting with high-purity specimens, the author has never observed the metastable state at $H_0 > H_{cr}$, so called "superheating" of the S phase, in spite of regular observation of the supercooling of the N phase. He was also not able to detect experimental reports on the superheating in literature.

Q vanishes. The same takes place at the lower end point of the curve (at $T = 0$ and $H_0 = H_{c0}$). The latter follows from the Third Law (Nernst's theorem), according to which at zero temperature $S_n = S_s = 0$. Therefore, somewhere at a mid temperature (see problem 3.1) the entropy difference $\Delta S(T) = S_n - S_s$ has a maximum and $\Delta C(T)$ passes through zero.

Also, in the upper end point, the vanishing of the latent heat is accompanied by the vanishing magnetic moment and by a jump of heat capacity. The latter, as follows from Eq. (3.9), is

$$\Delta C_c \equiv \Delta C(T \to T_c) = \frac{T_c V}{4\pi} \left(\frac{dH_{cr}}{dT} \right)^2. \qquad (3.10)$$

$$\Delta C_c \equiv \Delta C(T \to T_c) = \frac{T_c V}{\mu_0} \left(\frac{dB_{cr}}{dT} \right)^2.$$

This is the so-called Rutgers formula [83] presented by Rutgers' PhD adviser Paul Ehrenfest in discussion of his classification of phase transitions [84]. The Rutgers formula is consistent with experimental results obtained shortly beforehand [37] and was carefully verified thereafter (see [35] for references).

Hence, the discontinuous (first order) S/N transition below T_c degenerates into continuous (second order) phase transition in the close vicinity of the critical temperature. One cannot but notice a close similarity of properties of the S/N transition with those of the conventional liquid-gas transition.

However, in vicinity of the lower end point, this similarity is broken. Near the triple point solid-liquid-gas, densities and heat capacities of the coexisting phases are different and the latent heat of transition in either direction is not zero. In superconductors the magnetic moments of the coexisting phases are also different, but ΔS, Q and ΔC all vanish at approaching the lower end point. The S/N transition at $T \to 0$

is a quantum phase transition of the first order, similar to the melting transition in helium.

Completing this section, it is worth to stress that all that is said here was deduced from Eq. (3.6), reflecting the First Law, with the help of Eq. (1.32), following from the Poisson theorem, and of the Third Law. The only specific feature distinguishing the quantities involved in these calculations for superconductors from those for regular materials is presence of the condensation energy in Eq. (3.6).

3.2.2 Other geometries: Model of Peierls and London

In type-I superconductors of the cylindrical geometry, superconductivity is destroyed at once over the whole volume when the inner field H equals H_c and the outer field near the specimen surface $H_{ext}(= B_{ext})$ equals H_c as well. This indicates that the field H_c is the maximum field intensity inside the specimen at which superconductivity can survive in type-I materials. This can also be viewed as the fact that H_c is the maximum value of the induction of the external field above which superconductivity is destroyed. In the end, this fact can be interpreted as the pressure $H_c^2/8\pi$ (Problem 1.1) that the field B exerts on the S/N interface (a surface inside the specimen where B vanishes) is the maximum pressure which can be withstood by the S/N interface in type-I materials. All these interpretations are mutually related through the boundary condition (1.17). There was also a historically formed paradigm stating that at H_0 below H_c the N phase in type-I superconductors is unstable against transition to the S phase [66]. As seen from Fig. 3.1 and will be shown in more details below, this paradigm is strictly correct only for the cylindrical geometry.

Now we turn to the specimens of other than cylindrical geometry, i.e., to those with $\eta \neq 0$. The external field near such specimens in the Meissner state (Problem 1.2) is

$$H_{ext} = H_0 \frac{\sin \theta}{(1 - \eta)}. \qquad (1.38)$$

where θ is the angle which the normal to the surface makes with the applied field $\mathbf{H}_0$.

We see that H_{ext} vanishes near the "pole" (a line or point where $\theta = 0$) and reaches the maximum value $H_0/(1-\eta)$ near the "equator" (a line where $\theta = \pi/2$). $H_0/(1 - \eta)$ is also the magnitude of the field H inside the specimen (Problem 1.2). Hence, H becomes equal to H_c at $H_0 = (1 - \eta)H_c$. What happens at this field value?

On one hand, thermodynamics dictates that the specimen must remain superconducting. Indeed, from Eq. (3.4) we see that at $H_0 = (1 - \eta)H_c$ the specimen total free energy $\widetilde{F}$ is less than $\widetilde{F}_n$, meaning that there is still plenty of condensation energy available to support superconductivity. On the other hand, if the specimen continues staying in the Meissner state, then the condensation energy would be totally consumed and therefore the S/N transition would occur at $H_0 = H_c(1 - \eta)^{1/2}$. But if so, the specimen would be in the N state when the field (outside and inside) is uniform and less than H_c, which is impossible[7]. It is not difficult to figure out that a partial suppression of superconductivity at which the specimen core remains in the Meissner phase cannot happen either (see, e.g., [58]). Therefore, one can expect some drastic changes occurring at $H_0 = (1 - \eta)H_c$.

The first evidences of such changes were revealed in measurements of electrical resistivity in the magnetic field reported by de Haas with collaborators [85, 86]. It was found that the S/N transition in a tin rod, being very sharp when $\mathbf{H}_0$ is parallel to the rod longitudinal axis, is extended over about a half of the superconducting field range when $\mathbf{H}_0$ is perpendicular.

---

[7]From the phase diagram in Fig. 3.1, one can see that the N state below H_c is *possible* provided that the coexisting S state is the intermediate state.

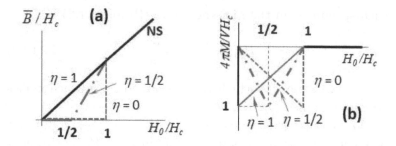

FIGURE 3.2 The average induction (a) and the magnetic moment (b), both in reduced units, in the Peierls-London model. Solid lines represent these quantities for specimens with demagnetizing factor $\eta = 1$ (infinite slab in perpendicular field), dash-dotted lines are for $\eta = 1/2$ (long circular cylinder in perpendicular field) and the dotted lines are for $\eta = 0$ (long cylinder in parallel field). NS designates the normal state (solid line at $H_0/H_c > 1$).

As mentioned in Chapter 1, at $H_0 > (1 - \eta)H_c$ the specimen is in the IS, where it splits for domains of the N phase, where $B = H$ or $\mu = 1$, and of the S phase, where $B = 0$ or $\mu = 0$. This entails that η is no longer defined and Eq. (1.22) is inapplicable.

The solution of the problem of magnetic properties of the IS was found independently by Peierls [57] and F. London [63]. They suggested to replace the inhomogeneous specimen by a homogeneous one with induction $\overline{B}$ averaged over the specimen volume. At the same time, the field intensity H throughout the specimen is assumed equal to its maximum value H_c reached at the critical field H_I of the transition from the Meissner state to the IS. In the Peierls-London (PL) model $H_I = (1 - \eta)H_c$. Hence, in this model the specimen is uniform both in the Meissner and in the intermediate states. Therefore, η is well defined and Eq. (1.22) is applicable all over the entire S state.

At the same time, the assumption $H = H_c$ makes the system of Equations (1.9) and (1.22) complete, thus allowing to calculate the average induction $\overline{B}$ and magnetization $\overline{I} = (\overline{B} - H_c)/4\pi$. Then the specimen magnetic moment is computed from definition $M = \overline{I}V$. It is important that M is directly measurable quantity.

Magnetic properties in the PL model are

$$H_0 \le H_c(1 - \eta) \begin{cases} B = 0 \\ H = H_0/(1 - \eta) \\ M = -H_0V/4\pi(1 - \eta) \end{cases} \quad (3.11)$$

$$B_0 \le B_c(1 - \eta) \begin{cases} B = 0 \\ H = B_0/\mu_0(1 - \eta) \\ M = -B_0V/\mu_0(1 - \eta) \end{cases} \quad \text{(SI)}$$

$$H_c(1 - \eta) \le H_0 \le H_c \begin{cases} \overline{B} = [H_0 - H_c(1 - \eta)]/\eta \\ H = H_c \\ M = -V(H_c - H_0)/4\pi\eta \end{cases} \quad (3.12)$$

$$B_c(1 - \eta) \le B_0 \le H_c \begin{cases} \overline{B} = [B_0 - B_c(1 - \eta)]/\eta \\ H = B_c/\mu_0 \\ M = -V(B_c - B_0)/\mu_0\eta \end{cases} \quad \text{(SI)}$$

Graphs for $\overline{B}$ and M are shown in reduced coordinates in Fig. 3.2. We see that the area under the graphs in Fig. 3.2b is $1/2$. Therefore, the PL model meets the mandatory requirement of thermodynamics. Another important feature is that it is applicable to specimens of all geometries (to ellipsoidal specimens of any shape), hence representing a global description of magnetic properties of type-I superconductors.

Experimental verification of the PL model revealed both its general consistency as well as some discordances with properties of real specimens (see, e.g., [35, 58] for references). In particular, in resistance [87] and magnetization [88] measurements on ZFC and FC cylindrical specimens (wires) in parallel and perpendicular fields it was found that H_I

is somewhat bigger than $H_c(1 - \eta)$ and the H_{cr} is somewhat less than H_c. Moreover, it turned out that differences $\Delta H_I = H_I - H_c(1 - \eta)$ and $\Delta H_{cr} = H_c - H_{cr}$ depend on the wire diameter: the smaller diameter, the bigger the difference. At the same time, the peak of magnetization curves, $4\pi M/V$ vs. H_0 for the transverse field was higher than H_c so that the areas under magnetization curves for the parallel and perpendicular fields are close to each other. The latter means that the experimental data meet the rule of $1/2$ and therefore these data reflect properties of the superconducting state.

A reason of such a behavior of magnetic properties of real specimens can be explained as follows.

One can show (see Problems 3.2 and 3.3) that the homogeneous specimen in the PL model is equivalent to its inhomogeneous analogy consisting of a set of laminae of the S and N phases with induction $B = 0$ and $B = H_c$, respectively. The laminae are parallel to each other and to the field $\mathbf{H} = \mathbf{H}_c$ inside the modeled specimen. The laminae are in a state of neutral static equilibrium, which implies that any parallel displacement, including aggregation of all same-type laminae together in one part of the sample, does not change the specimen's total free energy; in other words, the laminae do not interact and there is no surface barriers between them.

However, free energy of a real laminar structure, apart from the bulk contributions, which are accounted for in the PL model, contains surface contributions associated with (i) the S/N surface tension inside the specimens and (ii) an excess energy due to inhomogeneous field distribution and distortion of the domains' shape near the surfaces through which the flux enters and exits the specimen. The first of these two contributions favors a coarse domain structure (the less is the number of inter domain walls, the less the total energy of the interphase boundaries), whereas the second one favors a fine domain structure (the finer the structure, the smaller the surface related disturbances). Competition between these two terms optimizes and stabilizes the domain structure. This, in turn, affects all the rest equilibrium properties and the magnitude of these effects is determined by the ratio of the surface and the bulk energy contributions. Then one can expect that deviation of the properties of real specimens from those in the PL model decreases with increasing the specimen size, as it is observed in experiments. Naturally, none of these effects can affect the First Law, meaning that the rule of $1/2$ must be fulfilled for specimens of any geometry and any size.

Summarizing, the model of Peierls and London represents a global description of equilibrium magnetic properties of type-I superconductors of all (ellipsoidal) shapes in approximation of non-interacting laminae in the intermediate state (zero order approximation.) Applicability of this approximation is worsened with decreasing the specimen thickness[8]. The PL model provides a clear and simple picture of thermodynamic properties of the IS and serves as a general guide for understanding of magnetic properties of type-I superconductors. For that reason, importance of the PL model for superconductivity is similar to importance of the van der Waals model for phase transitions and critical phenomena.

3.2.3 Intermediate state in a flat slab: Laminar model for the slab in tilted field

A pioneering work demonstrating significance of the effects of the surface related inhomogeneities and of the S/N boundaries belongs to Landau [89]. Assuming that B at the S/N interface equals H_c and the Laplace equation for the scalar magnetic potential holds everywhere outside the specimen, Landau calculated the domains' shape and an excess energy of the disturbed field near the surface of a planar specimen in perpendicular field. Then taking into account the S/N surface tension, Landau calculated the period of the laminar structure and the width of the laminae.

---

[8]Dimension of the specimen in direction parallel to the field **H**.

Note that the first of Landau's assumptions is equivalent to the assumption $H = H_c$ in the PL model, hence automatically entails $H_{cr} = H_c$[9]. The second assumption implies that the field fills all the space outside the specimen. This resulted in the necessity to split the line of the external field at the middle of the S lamina and to bend it in the splitting point for 90 degrees. Apart from a potential conflict with the law of flux conservation, such a picture entails significant additional energy costs for maintaining the laminar structure, hence making the proposed model thermodynamically nonprofitable. Admitting this issue (indicated by Peierls) Landau abandoned his model and proposed so-called branching model instead [90, 91]. However, the latter was disproved in an experimental masterwork by Meshkovsky and Shalnikov developed according to Landau's recommendations [92]. Details of this dramatic story can be found in [11, 35, 54, 58].

There are three other possible scenarios of the surface related inhomogeneities in the IS [58]. The most appropriate one for modeling was proposed by Tinkham [54]. In this scenario, it is assumed that the dominant contribution in the surface related properties comes from field inhomogeneities extending over a "healing length" L_h outside the specimen, $L_h = (D_n^{-1} + D_s^{-1})^{-1}$, where D_n and D_s are the widths of the N and S laminae, respectively. Correspondingly, the roundness of the laminae corners is neglected. The field distribution in the Tinkham scenario is shown in Fig. 3.3. Advantages of this approximation are: (i) consistency with experimental images of the IS flux structure; (ii) absence of anchoring to any specific field, like H_c or $H_c(1 - \eta)$; (iii) meeting limiting cases for endpoints (when either D_n or D_s tend to zero, the healing length tends to zero as well); and last but not least (iv) simplicity.

---

[9]$B = H_c$ at the S/N interface means that $B = H = H_c$ over the whole volume of the N laminae since change of tangential component of B requires persistent current running in the N phase, which is impossible. On the other side, $B = H_c$ at the interface means that in the S laminae $H = H_c$ as well due to boundary condition (1.17).

The problem of the IS was revisited in [23, 34] via a set of experiments including magneto-optical imaging and resistance measurements in rotating field and measurements of magnetization in parallel and perpendicular fields performed with high purity indium films of different thicknesses. The most surprising result of this study is shown in Fig. 3.1: the critical field in the transverse field $H_{cr\perp}$ can be as small as about 40% of the thermodynamic critical field H_c[10]. As a result of this study, a new theoretical model of the IS was developed and validated experimentally. This model, which we will refer to as a laminar model for a flat slab in a tilted field (LMTF), is described below.

As it was for the first time shown by Sharvin [62] and confirmed in many experiments afterward [34, 93], the equilibrium flux structure of the IS in a flat slab subjected to a tilted field represents an ordered 1D lattice, like the one shown in Fig. 1.4b. Therefore the IS in such a specimen-field configuration is the most convenient for theoretical modeling. This is the reason for the geometry chosen in the LMTF model. Cross-section of the specimen in this model is shown in Fig. 3.3.

The model settings are the following:

(i) specimen is in the free space (vacuum).

(ii) specimen thickness $d \gg \lambda$. This means that negative surface tension of S/V (V stands for vacuum) interfaces due to non-zero $H_{0\|}$ is neglected.

(iii) longitudinal sizes of the specimen (along x and y axes) are much greater than thickness d, i.e., the slab is considered infinite. This means that flux of the perpendicular component of the applied field $H_{0\perp}$ is fixed and therefore $H_{0\perp} = \overline{B}_\perp = B_\perp \rho_n$, where $\overline{B}_\perp$ is average perpendicular component of the induction over the specimen; $B_\perp$ is perpendicular component of the induction in N domains; and ρ_n is volume fraction of the N phase: $\rho_n = D_n/D = V_n/V$

---

[10]In thinner films $H_{cr\perp}$ can be even less than that (V.K., not published).

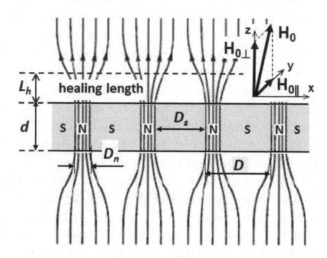

FIGURE 3.3 Cross-sectional view of the specimen/field configuration in the Tinkham scenario for a normally directed field
[54] and in the laminar model for the tilted field [34]. Specimen is an infinite plane-parallel slab. $\mathbf{H}_0$ is the applied field.
N and S are normal and superconducting domains represent-
ing rectangular parallelepipeds extended along $H_{0\parallel}$ component of the field (y-axis). D_n and D_s are widths of the N and
S domains, respectively. $D = D_n + D_s$ is the structure period
and d is the slab thickness. The healing length L_h is a characteristic distance over which the disturbed field relaxes to its
uniform distribution away from the specimen. In N domains
$B_\parallel = H_{0\parallel}$ and $B_\perp = H_{0\perp}/\rho_n$, where $\rho_n = D_n/D = V_n/V$
is the volume fraction of the N phase and V_n and V are
volumes of the N phase and of the entire specimen, respectively. (Reprinted from [23] with permission from the American Physical Society.)

with D_n and V_n designating the width of the N laminae and a total volume of the N phase, respectively.

(iv) $B_\| = H_\| = H_{0\|}$ due to absence of the demagnetizing field along the parallel component of the applied field $H_{0\|}$.

(v) Tinkham's version of the near-surface field distribution and domains' shape is adopted due to reasons indicated above.

Resolving $\mathbf{H}_0$ for components $H_{0\|}$ and $H_{0\perp}$, parallel and perpendicular to the specimen plane as shown in Fig. 3.3, we deal with the specimen of the cylindrical geometry for $H_{0\|}$ and of the transverse geometry for $H_{0\perp}$. Designating parts of the total free energy associated with these field components, correspondingly, as $\widetilde{F}_\|$ and $\widetilde{F}_\perp$ and referring to Table 2.1 we write

$$
\widetilde{F} = \widetilde{F}_\perp + \widetilde{F}_\| = V\left[\left(\widehat{f}_\| + \frac{H_{0\|}^2}{8\pi}\right) + \left(f_\perp - \frac{H_{0\perp}^2}{8\pi}\right)\right] =
$$

$$
V\left[\left(f_\| - \frac{B_\| H_\|}{4\pi} + \frac{H_{0\|}^2}{8\pi}\right) + \left(f_\perp - \frac{H_{0\perp}^2}{8\pi}\right)\right] =
$$

$$
V\left(f - \rho_n \frac{H_{0\|}^2}{4\pi} + \frac{H_{0\|}^2}{8\pi} - \frac{H_{0\perp}^2}{8\pi}\right), \quad (3.13)
$$

where we took into account that $B \neq 0$ only in N domains, where parallel component of the induction $B_\| = H_\| = H_{0\|}$.

The free energy $F = Vf$ consists of the following four components:

(i) the free energy at zero field

$$
V\left[f_n - (1 - \rho_n)\frac{H_c^2}{8\pi}\right], \quad (3.14)
$$

where we took into account that volume of the S phase V_s is $V(1 - \rho_n)$[11].

(ii) energy of the field $\mathbf{B}$ inside the N domains

$$
V\rho_n \frac{B_\|^2 + B_\perp^2}{8\pi} \quad (3.15)
$$

---

[11] At $H_0 = 0$, ρ_n vanishes and $V_s = V$.

(iii) energy of the S/N interphase walls inside the specimen (there are $2L_x/D$ walls with area $L_y d$ each)

$$2\frac{L_x}{D}(L_y d)\delta\frac{H_c^2}{8\pi} = 2V\frac{H_c^2\delta}{8\pi D}, \qquad (3.16)$$

where δ is the domain wall parameter discussed in Ch. 1 (see Fig. 1.3) and $\delta(H_c^2/8\pi)$ is the S/N interface energy per unite area of the wall.

(iv) an excess energy due to disturbed field over the healing length

$$\frac{2}{8\pi V}(L_h\rho_n B_\perp^2 L_x L_y - L_h H_{0\perp}^2 L_x L_y) = 2\frac{H_{0\perp}^2 D}{8\pi d}(1-\rho_n)^2. \qquad (3.17)$$

Now, we plug (3.14) through (3.17) into (3.13). Then the expression for the total free energy density takes the form

$$\tilde{f} = f_n - (1-\rho_n)\frac{H_c^2}{8\pi} + \frac{H_{0\perp}^2}{8\pi\rho_n} - \rho_n\frac{H_{0\parallel}^2}{8\pi} +$$

$$2\frac{H_c^2}{8\pi}\frac{\delta}{D} + 2\frac{H_{0\perp}^2}{8\pi}\frac{D}{d}(1-\rho_n)^2 + \frac{H_{0\parallel}^2}{8\pi} - \frac{H_{0\perp}^2}{8\pi}. \qquad (3.18)$$

Minimizing $\tilde{f}$ with respect to D, we find the optimal (i.e. equilibrium) period of the flux structure

$$D^2 = \frac{d\delta}{\rho_n^2(1-\rho_n)^2}\frac{H_c^2}{B_\perp^2} = \frac{d\delta}{(1-\rho_n)^2}\frac{H_c^2}{H_{0\perp}^2}. \qquad (3.19)$$

Plugging it in (3.18), we obtain

$$\tilde{f} = f_n - \frac{H_c^2}{8\pi}(1-\rho_n)\left[1 - h_{0\parallel}^2 - \frac{h_{0\perp}^2}{\rho_n} - 4h_{0\perp}\sqrt{\frac{\delta}{d}}\right], \qquad (3.20)$$

where $h_{0\parallel}$ and $h_{0\perp}$ are reduced components of the applied field $H_{0\parallel}/H_c$ and $H_{0\perp}/H_c$, respectively.

In SI units, this formula is

$$\tilde{f} = f_n - \frac{B_c^2}{2\mu_0}(1-\rho_n)\left[1 - b_{0\parallel}^2 - \frac{b_{0\perp}^2}{\rho_n} - 4b_{0\perp}\sqrt{\frac{\delta}{d}}\right], \qquad (\text{SI})$$

where $b_{0\parallel}$ and $b_{0\perp}$ are reduced components of the applied field $B_{0\parallel}/B_c$ and $B_{0\perp}/B_c$, respectively.

At this point, it is important to note that the term related to the S/N surface tension (3.16) and that related to the near-surface field disturbance (3.17) with the optimal D (Eq. (3.19)) are equal to each other. This implies that "responsibility" for deviation of the IS properties in this model from those in the PL model are equally shared between these two contributions. This is exactly what one should expect coming from the competitive nature of these contributions regardless of their specific mathematical form. Hence, D can be calculated simply by equating (3.16) and (3.17).

Minimizing (3.20) with respect to ρ_n (this is the only free parameter in the specimen's possession to optimize consumption of the condensation energy), we find the equilibrium volume fractions of the N and S components ρ_n and $\rho_s = 1 - \rho_n$, respectively

$$\rho_n^2 = \frac{h_{0\perp}^2}{1 - 4h_{0\perp}\sqrt{\frac{\delta}{d}} - h_{0\parallel}^2}. \tag{3.21}$$

Induction in the N laminae B (we do not use index since $B \neq 0$ only in the N phase) equals the field intensity H over the entire specimen[12]. Hence, in reduced units $b = B/H_c$ and $h = H/H_c$ ($b = B/B_c$ and $h = \mu_0 H/B_c$ in SI units) these quantities are

$$b^2 = h^2 = b_\perp^2 + b_\parallel^2 = h_{0\perp}^2/\rho_n^2 + h_{0\parallel}^2 = 1 - 4h_{0\perp}\sqrt{\frac{\delta}{d}} \tag{3.22}$$

$$b^2 = h^2 = b_\perp^2 + b_\parallel^2 = 1 - 4b_{0\perp}\sqrt{\frac{\delta}{d}}. \tag{SI}$$

---

[12]In the N laminae $H = B$ by definition of the N phase; in its turn, H in N and S laminae are equal due to the boundary condition (1-17).

Using $b_\perp$ from this formula $(b_\perp^2 = 1 - 4h_{0\perp}\sqrt{\delta/d} - h_{0\|}^2)$, Eq. (3.20) is transformed as

$$\widetilde{f} = f_n - \frac{H_c^2 b_\perp^2}{8\pi}(1 - \rho_n)^2 = f_n - \frac{H_c^2}{8\pi}(b_\perp - h_{0\perp})^2 \quad (3.23)$$

$$\widetilde{f} = f_n - \frac{B_c^2 b_\perp^2}{2\mu_0}(1 - \rho_n)^2 = f_n - \frac{B_c^2}{2\mu_0}(b_\perp - b_{0\perp})^2. \quad (\text{SI})$$

Now we calculate magnetic moment

$$\mathbf{M} \equiv -\nabla_{\mathbf{H_o}}(\widetilde{F}) = -\left(\frac{\partial \widetilde{F}}{\partial H_{0\|}}\hat{\mathbf{y}} + \frac{\partial \widetilde{F}}{\partial H_{0\perp}}\hat{\mathbf{z}}\right), \quad (3.24)$$

where $\hat{\mathbf{y}}$ and $\hat{\mathbf{z}}$ are unit vectors along the y and z axes, respectively.

Parallel component of $\mathbf{M}$ is

$$M_\| = -\frac{\partial \widetilde{F}}{\partial H_{0\|}} = -\frac{V}{H_c}\frac{\partial \widetilde{f}}{\partial h_{0\|}} = -\frac{V}{4\pi}(1 - \rho_n)H_{0\|} \quad (3.25)$$

$$M_\| = -\frac{\partial \widetilde{F}}{\partial B_{0\|}} = -\frac{V}{B_c}\frac{\partial \widetilde{f}}{\partial b_{0\|}} = -\frac{V}{\mu_0}(1 - \rho_n)B_{0\|}, \quad (\text{SI})$$

where we took into account that $\partial b_\perp/\partial h_{0\|} = -h_\|/b_\perp$ (see (Eq. (3.22)).

And the perpendicular component of $\mathbf{M}$ is

$$M_\perp = -\frac{\partial \widetilde{F}}{\partial H_{0\perp}} = -\frac{V}{H_c}\frac{\partial \widetilde{f}}{\partial h_{0\perp}} =$$

$$\frac{V}{4\pi}H_c(b_\perp - h_{0\perp}) \cdot \left(\frac{\partial b_\perp}{\partial h_{0\perp}} - 1\right) =$$

$$-\frac{V}{4\pi}(1 - \rho_n)\left(1 - \frac{\partial B_\perp}{\partial H_{0\perp}}\right)B_\perp. \quad (3.26)$$

$$M_\perp = -\frac{V}{\mu_0}(1 - \rho_n)\left(1 - \frac{\partial B_\perp}{\partial B_{0\perp}}\right)B_\perp. \quad (\text{SI})$$

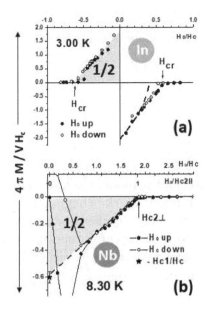

FIGURE 3.4 Typical magnetization curves for specimens of the transverse geometry of type-I (a) and type-II (b) superconductors reported in [23] and [47], respectively. Solid and open circles are data measured at ascending and descending applied field H_0. Initially the specimens were cooled at zero (i.e., Earth) field. The data in (a) were obtained on In film with thickness 3.86 μm. In (b) M was measured on a single crystal disc of 1 mm thick. Dashed curve in (a) represents $M(H_0)$ calculated from the LMTF. The dashed curve in (b) is $M(H)$ following from the AMMS. In both graphs 1/2 is area (shown in gray) under magnetization curve in reduced coordinates $4\pi M/V H_c$ vs H/H_c representing the condensation energy $H_c^2 V/8\pi$ with H_c inferred from the magnetization curves measured in parallel field. The star in (b) is magnetic moment following from the AMMS: $M(H_0 \to 0) = -H_{c1}V/4\pi$ with H_{c1} measured in parallel field, $H_{c2\parallel}$ and $H_{c2\perp}$ are the upper critical field, measured in parallel and perpendicular fields, respectively. (Reprinted from [23] and [47] with permissions from the American Physical Society and Springer Nature, respectively.)

At $H_0 = H_{cr}$ we have $\rho_n = 1$, $B_\perp = H_{0\perp}$, $\tilde{f} = f_n$ and $M = 0$. Using either of these conditions we find

$$h_{cr\perp} = \sqrt{\frac{4\delta}{d} + 1 - h_{0\|}^2} - 2\sqrt{\frac{\delta}{d}} \qquad (3.27)$$

$$b_{cr\perp} = \sqrt{\frac{4\delta}{d} + 1 - b_{0\|}^2} - 2\sqrt{\frac{\delta}{d}}. \qquad \text{(SI)} \qquad (3.28)$$

The first check of the model has to be on consistency with the rule of $1/2$. Integrating the components of $\mathbf{M}$ (3.25) and (3.26) back with respect to corresponding components of $\mathbf{H_0}$, one obtains magnetic energy of the specimen. Then the specimen total free energy takes the canonical form (3.4), implying that the model meets the rule of $1/2$ (see also Problem 3.5)

$$\tilde{F} = \tilde{F}(H_0 = 0) - \int_0^{H_0} \mathbf{M} \cdot d\mathbf{H_0} = F_{s0} - \int_0^{H_0} \mathbf{M} \cdot d\mathbf{H_0} =$$

$$F_n - \frac{H_c^2}{8\pi}V - \int_0^{H_0} \mathbf{M} \cdot d\mathbf{H_0}. \quad (3.29)$$

In parallel field ($H_{0\perp} = 0$), Eq. (3.21) yields $\rho_n = 0$, meaning that the specimen is in the Meissner state. Then, Eqs. (3.25) and (3.20) become as

$$M_\| = M = -\frac{V}{4\pi}H_0.$$

and

$$\tilde{f}_M = f_n - \frac{H_c^2}{8\pi} + \frac{H_0^2}{8\pi}$$

and therefore $H_{cr} = H_c$.

This is identical to the magnetic moment (see Fig. 1.2) and the total free energy Eq. (3.6) for specimens of the cylindrical geometry.

In perpendicular field ($H_{0\|} = 0$), we see that

(a) from (3.27): $H_{cr} < H_c$ in accord with experimental data [23, 34, 64, 87, 88];

(b) from (3.22): induction B starts from H_c at $H_0 \to$ $H_I = 0$ and decreases with increasing field down to H_{cr} at the S/N critical field. This is consistent with experimental data shown in Fig. 1.5;

(c) from (3.26): the magnetic moment at $H_0 \to 0 = H_I$ is

$$M(H_I) = -\frac{V}{4\pi} \left(1 - \frac{\partial B}{\partial H_0} \right) H_c.$$

Since B decreases with increasing H_0, magnitude of $4\pi M(H_I)/V$ (or $\mu_0 M(B_I)/V$ in SI units) is greater than its value in the PL model, where it is H_c (B_c). This is consistent with experimental data shown in Fig. 3.4a and reported in [23, 88].

At the same time from (3.27), (3.22) and (3.26) we see that deviations of H_{cr}, $B(H_0)$ and $M(H_I)$ from what follows from the PL model ($H_{cr} = H_c$, $H = H_c$ and $M(H_I) = H_c V/4\pi$) increases with decreasing the specimen thickness, in agreement with experiment [23, 86, 87, 88]. After all, for very thick specimens ($d \gg \delta$), the LMTF becomes identical to the PL model for $\eta = 1$ (see also Problem 3.2).

Quantitative comparison of the modeled and measured properties of the IS in the slabs was carried out in [23, 34]. It was found that some of the modeled properties fit experimental data nearly ideally, whereas other ones are less close. In particular, in the model ρ_n (Eq. (3.21)) and $M_\perp$ (Eq. (3.26)) are nonlinear functions of $H_{0\perp}$, whereas in experiments (e.g., [23, 34, 64, 87]) these functions are linear. An example of $M(H_0)$ data for type-I specimens of the transverse geometry is shown in Fig. 3.4a, where the magnetization curve following from the LMTF is shown by the dashed line. In [23, 34] deviation of the theoretical values from experimental data is attributed to the simplified approximation used to account the field distribution and domains' shape near the specimen surface.

3.3 TYPE-II SUPERCONDUCTORS: AVERAGED MODEL OF THE MIXED STATE

Fig. 3.5 presents a typical phase diagram of type II superconductors. Shown is the phase diagram constructed from magnetization data measured on Nb film specimen in parallel field
[47]. The thermodynamic critical field $H_c(T)$ was retrieved
from the area under measured magnetization curves. $H_c(T)$
characterizes density of the condensation energy and does not
depend on the field orientation. The latter is also true for the
curve of superconductivity nucleation $H_{c3}(T)$ [72]. One can
expect that the curve for the upper critical field $H_{cr}(T)(=
H_{c2}(T)$ when measured in the parallel field) depends on the
field orientation due to an excess energy associated with vortex ends; however, as of today no experimental observations
of the geometrical effects on $H_{cr}(T)$ were reported[13]. The
bottom curve is the curve of the transition from the Meissner state to the MS. For the given (cylindrical) geometry,
it is the low critical field $H_{c1}(T)$. For other geometries, it is
$H_{c1}(T)(1 - \eta)$; in specimens of the transverse geometry, the
bottom curve coincides with the T-axis. In high-purity specimens the critical fields H_{c1}, H_{c2}, H_{c3} and H_c are intrinsic
macroscopic parameters (dependent on temperature) of the
type-II material; otherwise, they are parameters of a specific
specimen. The critical fields are related: H_c is the geometric
mean of H_{c1} and H_{c2}, and $H_{c3} \approx 2H_{c2}$.

At $H_0 \leq H_{c1}(1 - \eta)$ a specimen of type-II superconductor
is in the Meissner state. Hence, it is magnetically homogeneous and therefore its demagnetizing factor is well defined.
Between $H_{c1}(1 - \eta)$ and H_{c2} the specimen is in the mixed
or Shubnikov state (MS), in which the flux passes through
in form of the single flux quantum vortices. Respectively, in
the MS η is not defined. Above $H_{c2}(T)$ there are traces of superconductivity in form of superconducting filaments, which

---

[13]This is attributed to the fact that the near-surface field distortions
due to single flux quantum vortices in the MS are much finer than the
distortions due to laminae in the IS [47].

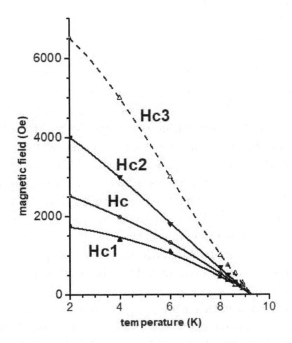

FIGURE 3.5   Phase diagram of type-II superconductors. Shown is the phase diagram of a 5.7-μm-thick high-purity niobium film specimen acquired from magnetization curves measured in parallel and perpendicular fields [47]. The field on the vertical axis is the applied field H_0 which at the measurements in parallel field equals the field intensity inside the specimen H. The curves $H_{c1}(T)$ and $H_{c2}(T)$ represent temperature dependencies of the lower and upper critical fields; $H_{c3}(T)$ is the critical field of nucleation of superconductivity; and $H_c(T)$ is the thermodynamic critical field retrieved from the area under the magnetization curves. (Reprinted from [47] with permission from Springer Nature.)

number density vanishes at H_{c3}. As was mentioned in Chapter 1, amount of the condensation energy involved in superconductivity above H_{c2} is about 1% of the total condensation energy $VH_c^2/8\pi$. On that ground, in further discussion, superconductivity above H_{c2} is neglected and the upper critical field H_{c2} is considered as the critical field of the S/N transition H_{cr} (with reservation that for thin specimens of non-cylindrical geometry H_{cr} may be lower than H_{c2}).

Thus, the main task of a theory of thermodynamic properties of type-II superconductors is the description of the MS properties. Most often, this task is pursued using the GL theory and the London theory modified to account for the flux passing through vortices which cores are assumed of a needle-like shape[14]. The theories are used to calculate average induction $\overline{B}$. Then magnetic moment is computed as $M = \overline{I}V$, where average magnetization $\overline{I} = (\overline{B}-H)/4\pi$. Both theories were used to calculate magnetic properties of specimens of cylindrical geometry for which $H(= H_0)$ is known [7, 12, 54, 56].

Hence, to use this approach for non-cylindrical geometries, one needs to know the field intensity H. This is the same problem, which was solved by Peierls and London for type-I specimens in the IS via assumption $H = H_c$. However, this assumption is inapplicable to type-II superconductors. On the other hand, if H in type-II specimens is known, it could allow one to construct a theoretical model for the MS similar to the PL model for the IS. Development of such a model requires (at least) availability of reversible experimental data on $M(H)$ for specimens of other than cylindrical geometry. In its turn, obtaining such data requires pinning free type-II specimens of appropriate size and shape, which, as it was mentioned in Chapter 1, is a much more challenging task than that for type-I materials.

---

[14]The modified London theory is supposed to be applicable to extreme type-II superconductors [53, 54]. However, even in this case validity of this theory is questionable [47].

A recent progress in fabrication of high purity Nb films [94] allowed to resolve the experimental part of the problem. The phase diagram in Fig. 3.5 is constructed from experimental magnetization curves obtained with one of these films.

Equilibrium magnetization curves for type-II superconductors of the transverse geometry were measured with film and single crystal Nb specimens in [47]. A typical result is shown in Fig. 3.4b. First we note that values of the critical field of the S/N transition (marked as $H_{c2\perp}$) are the same as those for the upper critical field H_{c2} measured in parallel field, marked as $H_{c2\parallel}$. The experimental data are reversible in about half of the field range of the S state. In the reversible part, the dependence M vs. H_0 is linear and its extrapolation (shown by the dashed line) yields $4\pi M/V$ close to H_{c1} at $H_0 \to 0$, where H_{c1} is the low critical field measured in parallel field. Validity of this extrapolation is supported by the rule of $1/2$: the area above the dashed line equals to the area above the $M(H_0)$ curve measured with the same specimen in the parallel field. It turned out that these experimental results are sufficient to find H in the MS and hence to construct a theoretical model of the MS magnetic properties. We will refer to this model as an averaged model of the mixed state (AMMS).

Similar to the PL model, in the AMMS the ellipsoidal specimen with spatially inhomogeneous induction B is replaced by a specimen of the same shape with an induction averaged over the volume $\overline{B}$. For such a specimen, η is well defined and Eq. (1.22) is valid. The rest directly follows from the experimental results on magnetization. That is how it goes [47].

The demagnetizing factor of a uniformly magnetized specimen of the transverse geometry η is unity and $\overline{B} = H_0$ due to the flux conservation. By definition, $H = \overline{B} - 4\pi M/V$[15]. Hence, linearity of $M(H_0)$ indicates that in the transverse geometry H is also a linear function of H_0. Then,

---

[15]We use H but not $\overline{H}$ because, same as in the IS, H is homogeneous throughout the specimen due to the boundary condition (1-17).

using experimentally found conditions $H_{c2\perp} = H_{c2\parallel}$ and $4\pi M(H_0 \to 0)/V = H_{c1}$, we obtain dependence $H(H_0)$ graphically shown by the line ab in Fig. 3.6b.

On the other hand, in specimens with $\eta = 0$ (cylindrical geometry) $H = H_0$ and therefore $H(H_0)$ is also a linear function connecting points $H = H_{c1}$ and $H = H_{c2}$. In Fig. 3.6b this is the dashed line between points c and b. Having that for the specimens of the cylindrical ($\eta = 0$) and transverse ($\eta = 1$) geometries the dependencies $H(H_0)$ are linear and extend from $H = H_{c1}$ to $H = H_{c2}$ one can conclude that for the rest geometries, i.e., for $0 < \eta < 1$, $H(H_0)$ is also linear and extends between the same values of H. In Fig. 3.6b $H(H_0)$ for $\eta = 1/2$ is shown by a dash-dotted line.

An analytical form of these functions is

$$H = H_{c1} + \frac{H_{c2} - H_{c1}}{H_{c2} - H_{c1}(1 - \eta)}[H_0 - H_{c1}(1 - \eta)]. \quad (3.30)$$

Thus, now we have two complete systems of equations:

(i) at $H_0 \leq H_{c1}(1 - \eta)$, the specimen is in the Meissner state and the equations are Eq. (1.9), Eq. (1.22) and $B = 0$.

(ii) at $H_0(1 - \eta) \leq H_0 \leq H_{c2}$, the specimen is in the MS and the system consists of Eq. (1.9), Eq. (1.22) and Eq. (3.30).

In the first field range, solutions are the same as in (3.11) with H_c replaced by H_{c1}.

In the second range $\overline{B}$ is

$$\overline{B} = \frac{H_{c2}}{H_{c2} - H_{c1}(1 - \eta)}[H_0 - H_{c1}(1 - \eta)] \quad (3.31)$$

$$\overline{B} = \frac{B_{c2}}{B_{c2} - B_{c1}(1 - \eta)}[B_0 - B_{c1}(1 - \eta)] \quad (SI) \quad (3.32)$$

and M is

$$\frac{4\pi M}{V} = -H_{c1} + \frac{H_{c1}}{H_{c2} - H_{c1}(1 - \eta)}[H_0 - H_{c1}(1 - \eta)] \quad (3.33)$$

$$\frac{\mu_0 M}{V} = -B_{c1} + \frac{B_{c1}}{B_{c2} - B_{c1}(1 - \eta)}[B_0 - B_{c1}(1 - \eta)]. \quad (SI)$$

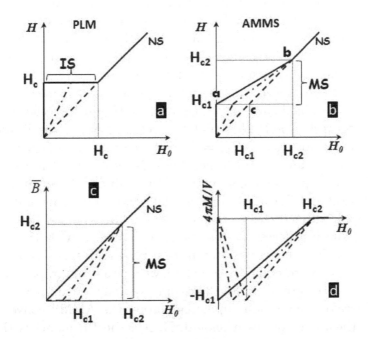

FIGURE 3.6 The field intensity H, average induction $\overline{B}$ and magnetic moment M vs applied field H_0 in specimens of different geometries. (a) H in the PL model for type-I superconductors; (b) H in the AMMS for type-II superconductors; (c) and (d) $\overline{B}$ and M (in the units of $4\pi M/V$) in the AMMS, respectively. Solid lines depict these quantities for specimens of the transverse geometry ($\eta = 1$); dash-dotted lines are for long circular cylinder in perpendicular field ($\eta = 1/2$), and the dashed lines are for specimens of cylindrical geometry ($\eta = 0$). IS, MS and NS stand for the intermediate, mixed and normal states, respectively. (Reprinted from [47] with permission from Springer Nature.)

Graphs for these quantities are shown in Figs. 3.6c and 3.6d.

We start testing the model from verification on compliance with the rule of $1/2$. From Fig. 3.6d we see that the area above the magnetization curves for different η is the same, implying that the condensation energy does not depend on the field orientation and therefore the model meets the rule of $1/2$ and, hence, the First Law of thermodynamics. One can also imaginally move H_{c2} toward H_{c1} in Fig. 3.6d, i.e., to convert type-II to type-I superconductor. In the latter, by definition $H_{c1} = H_{c2} = H_c$. We see that the magnetization curve for $\eta = 0$ (dashed lines) is becoming identical to that for type-I superconductor of the cylindrical geometry in Fig. 1.2 and Fig. 3.2b, whereas the section corresponding to the MS degenerates to the vertical line, as it should be.

Another necessary test is on compliance with the PL model. If we convert material from type-II to type-I moving point b toward point c in Fig. 3.6b, we see that the graphs convert to the graphs in Fig. 3.6a and Eq. (3.32) yields $H = H_c$. Therefore, for type-I superconductors, the AMMS converts to the PL model as it should. Hence, the PL model is the limiting case of the AMMS for type-I superconductors.

From the graphs in Fig. 3.6d and Eq. (1.37), it follows that

$$-\int_0^{H_{c2}} \frac{4\pi \mathbf{M}}{V} d\mathbf{H}_0 = \frac{H_{c1} H_{c2}}{2} = \frac{H_c^2}{2}.$$

Hence, according to the model, H_c is a geometrical mean of H_{c1} and H_{c2}. This is an empirical rule approximately justified[16] for extreme type-II superconductors [54]. The AMMS extends this rule to all type-II materials.

The dashed line in Fig. 3.4b represents the magnetization curve in AMMS for the transverse geometry. We see that the model quantitatively matches the experimental data for this geometry. However, consistency with the experiment worsens with decreasing η. Experimental magnetization curve (see,

---

[16]In the GL theory $H_{c1} H_{c2} = H_c^2 \ln \kappa$ [56].

e.g., Fig. 1.5) in the MS is not linear, whereas in the model $M(H_0)$ in the MS is the straight line. In [47] this discrepancy is attributed to the surface current present in specimens of cylindrical geometry. Also, one cannot exclude a possibility of surface barriers in real specimens of the non-transverse geometry [95]. Since the specimen of the transverse geometry does not have the side surface, the surface effects in such a specimen are absent by definition. The latter implies that equilibrium magnetic properties of a specimen of transverse geometry represent intrinsic properties of the vortex ensemble (vortex matter).

Detailed discussion of the AMMS is available in [47]. There it is shown that, like the PL model for type-I superconductors, the AMMS describes equilibrium magnetic properties of type-II superconductors of all shapes and all κ in the limit of non-interacting vortices. The model accounts for the negative S/N surface tension via introduction of $H_{c1} < H_c < H_{c2}$[17], but neglects the effects related to the vortices' ends and of the lateral surface.

We complete this last section by noting that the quantitative agreement of the modeled and experimental magnetization curves for the transverse geometry provides a solid commencement for a number of fundamental conclusions about the vortex matter discussed in [47]. In particular, that a frequently used concept of vortex-vortex interaction in the equilibrium state is physically unjustified. However, properties of both individual vortices and the vortex matter is a separate topic which goes beyond the scope of this book.

3.4 PROBLEMS

3.1. (a) At what temperature T_1 the difference of entropies in the N and the S state of type-I superconductors of the cylindrical geometry is maximal? (b) What happens to the difference of heat capacities ΔC at this temperature? (c) How

---

[17]In absence of the *negative* surface tension, $H_{c1} = H_{c2}$.

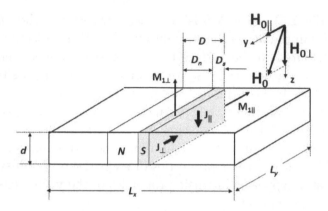

FIGURE 3.7 Schematics of the specimen in Problems 3.2-3.4. N and S mark the normal and superconducting laminae assumed of the shape of rectangular parallelepipeds with widths D_n and D_s, respectively. $D = D_n + D_s$ is period of the flux structure. $J_\perp$ and $J_\parallel$ are surface currents in S lamina due to $H_{0\perp}$ and $H_{0\parallel}$, respectively. $M_{1\perp}$ and $M_{1\parallel}$ are magnetic moments of a single S domain due to perpendicular and parallel components of the applied field, respectively.

does ΔC depend on T below T_1? Use the empirical expression (3.2) for $H_c(T)$.

Answer: (a) Entropy difference $\Delta S = S_n - S_s$ passes through maximum at $T_1 = T_c/\sqrt{3}$; (b) difference of heat capacities $\Delta C = C_s - C_n$ at T_1 is zero; (c) At $T < T_1$ ΔC is negative; since $\Delta C = 0$ at $T = 0$ (the Third Law), below T_1 ΔC passes though minimum. Using (3.2) for $H_c(T)$ one finds that minimum of ΔC takes place at $T_2 = T_c/\sqrt{5}$.

3.2. Consider a superconducting slab in perpendicular field $H_0 = H_{0\perp}$ depicted in Fig. 3.7. Dimensions of the slab $L_x \sim L_y \gg d$. Assume that flux passing the slab forms a 1D lattice with period D of the N and S domains which widths are D_n and D_s, respectively. Induction B in N and S domains equals H_c and zero, respectively. Neglect the near-surface distortions of the field and domains' shape.

Show that magnetic properties of this slab ($\mathbf{M}$, $\mathbf{H}$ and $\mathbf{B}$) are identical to those of a specimen with $\eta = 1$ in the PL model.

Solution.

Referring to Fig. 3.7 with $H_{0\|} = 0$ and using (1.18), we write

$$M_1 = \frac{(L_y D_s) J_\perp}{C} = -\frac{L_y D_s}{c} \frac{c(B - 0)}{4\pi} d = -\frac{L_y D_s}{c} \frac{c H_c}{4\pi} d.$$

$$M = \frac{L_x}{D} M_1 = -\frac{H_c}{4\pi} (L_x L_y d) \frac{D_s}{D} = -\frac{H_c}{4\pi} V \rho_s$$

$$\rho_s \equiv \frac{V_s}{V} = \frac{D_s}{D} = 1 - \rho_n = 1 - \frac{D_n}{D} = 1 - \frac{H_0}{H_c},$$

where the last expression follows from the flux conservation.

So,

$$M = -\frac{V H_c}{4\pi} (1 - \rho_n) = -\frac{V}{4\pi} (H_c - H_0).$$

Note that M does not depend on specific flux structure, but only on the total volume of the N phase $V_n = \rho_n V$; hence, one is free to relocate domains keeping constant V_n. This implies that the domains can be either ordered, like in Fig. 3.7, or disordered, like in Fig. 1.4a, or they can be collected all together in one place. The magnetic moment of all these systems will be the same as soon as energy contribution due to surface-related disturbances along with the contribution due to the S/N boundaries can be neglected. According to the LMTF, this takes place in sufficiently thick specimens, i.e., in specimens with $d \gg \delta$.

Now, due to the boundary condition (1.17)

$$H = H_c.$$

And from (1.9)

$$\overline{B} = H + \frac{4\pi M}{V} = H_0.$$

$\mathbf{M}$, $\mathbf{H}$ and $\mathbf{B}$ are the same as in the PL model for $\eta = 1$ (Eqs. (3.12)). Q.E.D.

3.3 For the specimen of the previous problem, compose expression for the total free energy $\widetilde{F}(H_0)$ and calculate M.
Solution.
Referring to Table 2.1 and Fig. 3.7 and accounting that $\widetilde{F}_{n0} = F_{n0}$, we write

$$\widetilde{F} = F - \frac{H_0^2}{8\pi}V =$$

$$F_{n0} - \left(\frac{H_c^2}{8\pi}dL_yD_s\right)N + \left(\frac{H_c^2}{8\pi}dL_yD_n\right)N - \frac{H_0^2}{8\pi}V =$$

$$F_{n0} - \frac{H_c^2}{8\pi}V\rho_s + \frac{H_c^2}{8\pi}V\rho_n - \frac{H_0^2}{8\pi}V.$$

Taking into account that $\rho_n = H_0/H_c$ and $\rho_s = 1 - \rho_n$ we continue

$$\widetilde{F} = F_{n0} + \frac{VH_c^2}{8\pi}(2\rho_n - 1) - \frac{H_0^2}{8\pi}V =$$

$$F_{n0} + \frac{VH_c^2}{8\pi}\left(2\frac{H_0}{H_c} - 1\right) - \frac{H_0^2}{8\pi}V = F_{n0} =$$

$$F_{n0} - \frac{V}{8\pi}(H_c - H_0)^2.$$

Note that $\widetilde{F}$ in this model does not contain a free parameter which could be varied to minimize the free energy, i.e., to optimize consumption of the condensation energy. In other words, any flux pattern (if look from the top in Fig. 3.7) has the same free energy and therefore equally probable, meaning that the system is in the neutral static equilibrium.
The magnetic moment of this specimen is

$$M = -\frac{\partial \widetilde{F}}{\partial H_0} = -\frac{V}{4\pi}(H_c - H_0),$$

M is the same as in the previous problem and in the PL model for $\eta = 1$, as it should be.

3.4. The slab schematically shown in Fig. 3.7 is in a tilted field (both $H_{0\perp}$ and $H_{0\|}$ are not zero). As in the previous two problems, induction in N and S domains is H_c and 0, respectively. Calculate M from magnetostatics (using surface currents $J_\perp$ and $J_\|$). Is **M** aligned with **H**$_0$?

Hint. Brake **B** in N domain to $B_\perp$ and $B_\|$ components; use the flux conservation law $B_\perp = H_{0\perp}/\rho_n$ and the boundary condition (1.17) for $B_\|(= H_{0\|})$. Then performing the same steps as in Problem 3.2, calculate $M_\perp$ and $M_\|$.

Answer.

$$M_\perp = -\frac{B_\perp V}{4\pi}(1-\rho_n) = -\frac{V}{4\pi}(B_\perp - H_{0\perp}).$$

$$M_\| = -\frac{B_\| V}{4\pi}(1-\rho_n) = -\frac{V}{4\pi}(B_\perp - H_{0\perp})\frac{H_{0\|}}{B_\perp}.$$

$$M = -\frac{V}{4\pi}(1-\rho_n)H_c$$

in the vector form

$$\mathbf{M} = -\frac{V}{4\pi}(1-\rho_n)\mathbf{B}$$

with

$$\mathbf{B} = \frac{H_{0\perp}}{\rho_n}\hat{\mathbf{z}} + H_{0\|}\hat{\mathbf{y}}.$$

where $\hat{\mathbf{z}}$ and $\hat{\mathbf{y}}$ are unit vectors along z and y axes.

M is antiparallel to (aligned with) **B** in the N domains, i.e., (due to the boundary condition (1.17)) to **H**, but not to **H**$_0$. Direction of **B** and **H** change with changing *magnitude* of **H**$_0$.

3.5. Consider a specimen of the transverse geometry in the PL model. (a) What is entropy S_s and heat capacity C_s of this specimen? (b) What are the differences $\Delta S = S_s - S_n$ and $\Delta C = C_s - C_n$ along the coexisting curve $H_{cr}(T)$[18]? (c) What kind of the phase transition is the S/N transition in this model?

---

[18] S_n and C_n are entropy and heat capacity in the N state.

Solution.

In the PL model for $\eta = 1$: $H = H_c$, $\overline{B} = H_0$, $H_{cr} = H_c$ and $M = -V(H_c - H_0)/4\pi$. So

$$\widetilde{F} = F_{n0} - \frac{V}{8\pi}(H_c - H_0)^2.$$

(a) In the superconducting (in this case intermediate) state

$$S_s = -\left(\frac{\partial \widetilde{F}}{\partial T}\right)_{H_0} = S_n + \frac{V}{4\pi}(H_c - H_0)\frac{dH_c}{dT}.$$

And heat capacity C_s (this is C_s at constant $\overline{B}$ since in this geometry we control $\overline{B} = H_0$) is

$$C_s = T\left(\frac{\partial S_s}{\partial T}\right)_{H_0} = C_n + \frac{VT}{4\pi}\left[\left(\frac{dH_c}{dT}\right)^2 + (H_c - H_0)\frac{d^2H_c}{dT^2}\right].$$

(b) Along the coexisting curve $H_0 = H_{cr}(T) = H_c(T)$

$$\Delta S = S_s - S_n = 0 \qquad (3.34)$$

and

$$\Delta C = C_s - C_n = \frac{VT}{4\pi}\left(\frac{dH_c}{dT}\right)^2 > 0.$$

(c) In the PL model for specimens with $\eta = 1$ the S/N transition is thermodynamic phase transition of the second order. The same is true for the IS/N transition in a specimen of any other geometry ($0 < \eta \leq 1$) within the PL model. The reason of that is neglect of the barriers between the laminae in this model.

3.6. Find magnetic energy E_m (defined by Eq. (1.25)) of the specimen in the AMMS. Show that $E_m = H_c^2 V/8\pi$ at $H_0 = H_{cr}$.

Solution. Referring to (3.23), we write the total free energy density as

$$\tilde{f} = f_n - \frac{H_c}{8\pi} + \frac{H_c^2}{8\pi} \left[\rho_n + (1 - \rho_n) \left(h_{0\parallel}^2 + \frac{h_{0\perp}^2}{\rho_n} + 4h_{0\perp}\sqrt{\frac{\delta}{d}} \right) \right],$$

Comparing with (3.4), we write,

$$E_m = \frac{H_c^2 V}{8\pi} \left[\rho_n + (1 - \rho_n) \left(h_{0\parallel}^2 + \frac{h_{0\perp}^2}{\rho_n} + 4h_{0\perp}\sqrt{\frac{\delta}{d}} \right) \right].$$

At $H_0 = H_{cr}$ the fraction $\rho_n = 1$ and therefore $E_m = H_c^2 V/8\pi$. Q.E.D.

3.7.

(a). Compose an "equation of state" (number density n_v vs. applied field H_0) of an equilibrium vortex assemble ("vortex matter") in specimens of the transverse geometry. (b) Find "compressibility" $\beta = n_v^{-1} \partial n_v / \partial H_0$ of this matter. What is its (c) entropy S_v (a measure of the vortex disorder) and (d) "temperature" T_v (a measure of kinetic energy of vortices)?

Solution.

An ensemble of single flux quantum vortices in type-II superconductors is referred to as the vortex matter, where the roles of pressure and mass density are taken by the applied field H_0 and the vortex number density n_v (the number of vortices per unite area of a plane perpendicular to the field), respectively. Equation of state of the vortex matter is dependence of n_v vs. H_0.

Magnetic properties of pinning-free type-II superconductors of the transverse geometry represent equilibrium properties of pure vortex matter in view of the absence of effects associated with the side surface and pinning. In this geometry the average induction $\overline{B} = H_0$ and the total flux is $B_0 A = H_0 A = \Phi_0 N$, where A is the specimen area and N is the number of all vortices. Hence,

(a) $n_v \equiv N/A = H_0/\Phi_0$.

(b) $\beta = 1/H_0$.

(c) $S_v = 0$ since vortices are aligned with the field and are ordered into the hexagonal lattice in the plane perpendicular to the field.

(d) $T_v = 0$ due to the Third law. This means that, unlike molecules in a regular matter at constant pressure, vortices at the fixed H_0 are motionless.

Bibliography

[1] Grigoriev I. S., Meilikhov E. Z., Radzig A. A., Eds., *Handbook of Physical Quantities*, (CRC Press, Boca Raton, 1997).

[2] M. Faraday, *Experimental Researches in Electricity* v.III, (Taylor & Francis, London, 1855).

[3] I. E. Tamm, *Fundamentals of the Theory of Electricity*, (Mir, Moscow, 1979).

[4] J. D. Jackson, The nature of intrinsic magnetic dipole moments (CERN-77-17). European Organization for Nuclear Research (CERN), 1977.

[5] L. D. Landau and E. M. Lifschitz, *The Classical Theory of Fields*, 4th ed. (Elsevier, 1975).

[6] J. D. Jackson, *Classical Electrodynamics*, 3d ed. (John Wiley & Sons, Inc., Hoboken NJ, 1999).

[7] L. D. Landau, E. M. Lifshitz and L. P. Pitaevskii, *Electrodynamics of Continuous Media*, 2nd ed. (Elsevier, 1984).

[8] D. J. Griffiths, *Introduction to Electrodynamics*, 4th ed. (Pearson, Edinburgh, 2014)

[9] H. A. Lorentz, *The Theory of Electrons* (The Columbia University Press, New York, 1909).

[10] J. C. Maxwell, *A Treatise on Electricity and Magnetism*, v.II (Clarendon Press, Oxford, 1873).

[11] P. G. de Gennes, *Superconductivity of Metals and Alloys* (Perseus Book Publishing, L.L.C., 1966).

[12] A. A. Abrikosov, *Fundamentals of the Theory of Metals* (Elsevier Science Pub. Co., 1988).

[13] E. M. Purcell, *Electricity and Magnetism*, 2nd ed. (McGraw Hill, Boston, 1985).

[14] Par M. Poisson, Memoire sur La Theorie Du Magnetisme, Lu a l'Academie Royale des Sciences, 2 Fevrier, 1824.

[15] E. A. Guggenheim, Proc. Roy. Soc. **115A**, 49 (1936).

[16] C. Kittel, Introduction to Solid State Physics, 8th ed. (John Wiley & Sons, Inc., NJ, 2005).

[17] R.P. Lungu, in *Trends in Electromagnetism*, Ed. V. Barsan, p. 113 (Intechopen, Rijeka 2012).

[18] V. Franco, J. S. Blázquez, J. J. Ipus, J. Y. Law, L. M. Moreno-Ramírez, A. Conde, Progr. Materials Science **93**, 112 (2018).

[19] H.-J. Van Leeuwen, J. Phys. Radium **2**, 361 (1921).

[20] J. H. Van Vleck, *The Theory of Electric and Magnetic Susceptibilities* (Clarendon Press, Oxford, 1932).

[21] F. Bloch, *Molekulartheorie des Magnetismus* (Akad. Verlagsgesellschaft, Leipzig, 1934).

[22] R. Feynman, R. Leighton, M. Sands, The Feynman Lectures on Physics, v. II (Basic Books, N.Y., 1964).

[23] V. Kozhevnikov and C. Van Haesendonck, Phys. Rev. B **90**, 104519 (2014).

[24] N. W. Ashcroft and N. D. Mermin, Solid State Physics (Brooks/Cole Cengage Learning. Belmont, USA, 1976).

[25] *Perspectives in Quantum Hall Effects*, S. Das Sarma and A. Pinczuk (Eds), (Wiley, New York, 1997).

[26] L. D. Landau and E. M. Lifshitz, *Statistical Physics*, Part I, 3d ed. (Elsevier, Amsterdam, 1980).

[27] H. Kamerlingh Onnes, Comm. Phys. Lab. Univ. Leiden, No. 124c (1911).

[28] B. T. Matthias, T. H. Geballe, and V. B. Compton, Rev. Mod. Phys. **35**, 414 (1963).

[29] B. T. Matthias, Science **144**, 378 (1964).

[30] Novel Superconductivity, Eds. V. Z. Kresin and S. A. Wolf (Premium Press, N.Y., 1987).

[31] Rogalls H. and Kes P. H., Eds., *100 Years of Superconductivity*, (CRC Press, Boca Raton, 2012).

[32] H. Kamerlingh Onnes, Comm. Phys. Lab. Univ. Leiden, Nos. 104b, 140c (1914).

[33] A. B. Pippard, *The Dynamics of Conduction Electrons* (Gordon and Breach, New York, 1965).

[34] V. Kozhevnikov, R. J. Wijngaarden, J. de Wit, and C. Van Haesendonck. PRB 89, 100503(R) (2014).

[35] D. Shoenberg, *Superconductivity*, 2nd. ed., (Cambridge University Press, 1952).

[36] F. London, *Superfluids* v. 1, 2nd ed. (Dover, N,Y., 1961.)

[37] W. H. Keesom and J. N. van den Ende, Comm. Phys. Lab. Univ. Leiden, No. 219b (1932); W. H. Keesom and J. A. Kok, Comm. Phys. Lab. Univ. Leiden, No. 221b (1932); Proc. Ken. Akad. Amsterdam **35**, 743 (1932).

[38] J. Mehra, *The Solvay Conferences on Physics* Ch. 5 (D. Reidel Publishing Co., Dordrecht-Holand, 1975).

[39] W. Meissner and R. Ochsenfeld, Die Naturwissenschaften **21** 787 (1933); English translation is available at A. M Forrest, Eur. J. Phys. **4**, 117 (1983).

[40] G. N. Rjabinin and L. W. Shubnikow, Nature **134**, 286 (1934).

[41] P. F. Dahl, Hist. Stud. Phys. Biol. Sci. **16**, 1 (1986).

[42] A. B. Pippard, *Elements of Classical Thermodynamics for Advanced Students of Physics* (Cambridge University Press, London, 1966).

[43] J. Bardeen, L. N. Cooper and J. R. Schrieffer, Phys. Rev. **108**, 1175 (1957).

[44] J. R. Schrieffer, *Theory of Superconductivity* (Perseus Books, Reading, 1999).

[45] N. N. Bogoliubov, Sov. Phys. JETP **34**, 41 (1958); V. V. Tolmachev and S. V. Tiablikov, *iden* **34**, 46 (1958); N. N. Bogoliubov *iden* **34**, 51 (1958).

[46] N. N. Bogolyubov, V. V. Tolmachev, D. N. Shirkov, *A New method in the Theory of Superconductivity* (Consultants Bureau, New York, 1959).

[47] V. Kozhevnikov, A.-M. Feliciano, P. J. Curran, G. Richter, A. Volodin, A. Suter, S. J. Bending, C.Van Haesendonck, J Supercond. Nov. Magn. **31**, 3433 (2018).

[48] V. L. Ginzburg and L. D. Landau, Zh.E.T.F. **20**, 1064 (1950).

[49] A. B. Pippard, Proc. Cambridge Phil. Soc. **47**, 617 (1951).

[50] F. and H. London, Physica **2**, 341 (1935).

[51] C. J. Gorter and H. B. G. Casimir, Phys. Z. 35, 963 (1934).

[52] R. P. Feynman, in *Progress in Low temperature Physics* Gorter, C. J. (ed.) vol. I, pp. 17-53. (North Holland Publishing Company, Amsterdam, 1955).

[53] E. M. Lifshitz and L. P. Pitaevskii *Statistical Physics* v.2, (M., Nauka, 1973).

[54] M. Tinkham, *Introduction to Superconductivity*, 2nd ed. (McGraw-Hill, 1996).

[55] L. P. Gorkov, in 100 Years of Superconductivity, p. 72, Eds. H. Rogalla and P. H. Kes, (CRC Press, Boca Raton, 2012).

[56] A. A. Abrikosov, Zh.E.T.F. 32, 1442 (1957).

[57] R. Peierls, Proc. R. Soc. London, Ser. A **155**, 613 (1936).

[58] V. Kozhevnikov, in *Superfluids and Superconductors*, Ed. R. Zivieri (IntechOpen, London, 2018).

[59] C. J. Gorter and H. Casimir, Physica **1**, 306 (1934).

[60] A. I. Shalnikov, Zh.E.T.F. **33**, 1071 (1957).

[61] I. T. Faber, Proc. Roy. Soc. A **248**, 460 (1958).

[62] Yu. V. Sharvin, Zh.E.T.F. **33**, 1341 (1957)

[63] F. London, Physica **3**, 450 (1936).

[64] V. S. Egorov, G. Solt, C. Baines, D. Herlach, and U. Zimmermann, Phys. Rev. B **64**, 024524 (2001).

[65] L.V. Shubnikov, V.I. Khotkevich, Yu.D. Shepelev, Yu.N. Ryabinin, Zh.E.T.F. **7**, 221 (1937).

[66] H. London, Proc. Roy Soc. Lond. A **152**, 650 (1935).

[67] U. Essmann and H. Träuble, Phys. Lett. **24A**, 526 (1967).

[68] A. Volodin, C. Van Haesendonck, Physicalia Mag. **26**, 141, (2004).

[69] L. Embon, Y. Anahory, Z. L. Jelic, E. O. Lachman, Y. Myasoedov, M. E. Huber, G. P. Mikitik, A. V. Silhanek, M. V. Milosevic, A. Gurevich and E. Zeldov, Nature Communication **8**, 85 (2017).

[70] D. K. Finnemore, T. F. Stronberg and C. A. Swenson, Phys. Rev. **149**, 231 (1966).

[71] D. Saint-James and P. G. De Gennes, Phys. Letters **7**, 306 (1963).

[72] V. Kozhevnikov, A.-M. Valente-Feliciano, P. J. Curran, A. Suter, A. H. Liu, G. Richter, E. Morenzoni, S. J. Bending, and C. Van Haesendonck, Phys. Rev. B **95**, 174509 (2017).

[73] L. P. Gorkov and V. Z. Kresin, Rev. Mod. Phys. **90**, 011001 (2018).

[74] A. P. Drozdov, P. P. Kong, V. S. Minkov, S. P. Besedin, M. A. Kuzovnikov, S. Mozaffari, L. Balicas, F. Balakirev, D. Graf, V. B. Prakapenka, E. Greenberg, D. A. Knyazev, M. Tkacz, M. I. Eremets, arXiv:1812.01561 (2018).

[75] E. A. Guggenheim, Proc. Roy. Soc. **115A**, 70 (1936).

[76] C. Kittel and H. Kroemer, *Thermal physics*, 2nd ed. (Freeman, N.Y., 1980)

[77] D. K. Finnemore and D. E. Mapother, Phys. Rev. **140**, A507 (1965).

[78] V. Kozhevnikov, A. Suter, H. Fritzsche, V. Gladilin, A. Volodin, T. Moorkens, M. Trekels, J. Cuppens, B. Wojek, T. Prokscha, E. Morenzoni, G. J. Nieuwenhuys, M. J. Van Bael, K. Temst, C. Van Haesendonck, and J. O. Indekeu Phys. Rev. B **87**, 104508 (2013).

[79] A. Suter, E. Morenzoni, N. Garifianov, R. Khasanov, E. Kirk, H. Luetkens, T. Prokscha, and M. Horisberger, Phys. Rev. **B 72**, 024506 (2005).

[80] R. F. Kiefl, M. D. Hossain, B. M. Wojek, S. R. Dunsiger, G. D. Morris, T. Prokscha, Z. Salman, J. Baglo, D. A. Bonn, R. Liang, W. N. Hardy, A. Suter, and E. Morenzoni, Phys. Rev. B **81**, 180502(R) (2010).

[81] Oren Ofer, J. C. Baglo, M. D. Hossain, R. F. Kiefl, W. N. Hardy, A. Thaler, H. Kim, M. A. Tanatar, P. C. Canfield, R. Prozorov, G. M. Luke, E. Morenzoni, H. Saadaoui, A. Suter, T. Prokscha, B. M. Wojek, and Z. Salman, Phys. Rev. B **85**, 060506(R) (2012).

[82] W. H. Keesom, in Rapports et Discussions du Quatrieme Conseil de Physique, (Solvay Institut, Paris, 1927), p. 288 (Proc. 4th Solvay Conference of 1924. Comment of Keesom at discussion of a report of Kamerlingh-Onnes.)

[83] A. J. Rutgers, Physica **1**, 1055 (1934).

[84] P. Ehrenfest, Comm. Leiden Suppl. No. 75b (1933).

[85] W. J. de Haas and J. Voogd, Commun. Phys. Lab. Leiden, No 212d; KNAW Proc. **34**(1), 63 (1931).

[86] W. J. de Haas, J. Voogd, J.M. Jonker, Physica **I**, 281 (1934).

[87] E.R. Andrew, Proc. R. Soc. Lond. A **194**, 80 (1948).

[88] M. Desirant and D. Shoenberg, Proc. Roy. Soc. London, A194, 63 (1948).

[89] L. D. Landau, Zh.E.T.F. **7**, 371 (1937).

[90] L. D. Landau, Nature **141**, 688 (1938).

[91] L. D. Landau, Zh.E.T.F. **13**, 377 (1943).

[92] A. G. Meshkovsky and A. I. Shalnikov, Zh.E.T.F. **17**, 851 (1947).

[93] R. P. Huebener, *Magnetic Flux Structures in Superconductors*, 2nd Ed. (Springler-Verlag, N.Y., 2010).

[94] A.-M. Valente-Feliciano, Development of SRF mono-layer/multilayer thin film materials to increase the performance of SRF accelerating structures beyond bulk Nb, PhD dissertation, Université Paris Sud - Paris XI, 2014.

[95] C. P. Bean and J. D. Livingston, Phys. Rev. Letters **12**, 14 (1964).

Index

9 780367 788018